THE BIG BOOK OF
SCIENCE
& TECHNOLOGY

ROBIN KERROD

SMITHMARK

This edition published in 1991 by SMITHMARK Publishers Inc.,
112 Madison Avenue, New York, NY 10016
By arrangement with Reed International Books
Michelin House, 81 Fulham Road, London SW3 6RB

ISBN 0-8317-0853-0

Printed in Italy

SMITHMARK Books are available for bulk purchase
for sales promotion and premium use.
For details write or telephone the Manager of Special Sales,
SMITHMARK Publishers Inc., 112 Madison Avenue,
New York, NY 10016. (212) 532-6600

CONTENTS

EXPLORING MATTER
Matter and the Universe   8
Crystals   10
Liquids   12
Looking at Gases   14
Reaction and Change   16
Building Blocks of Matter   18
Inside Atoms   20

EXPLORING MATERIALS
Raw Materials   22
Metals and their Uses   24
Oil Distillation   26
Versatile Plastics   28
Pesticides and Drugs   30

EXPLORING FORCES
Movement   32
Falling Things   34
Beating Gravity   36
Magnetism   38
Simple Machines   40

EXPLORING ENERGY
Hot and Cold   42
Fossil Fuels   44
Energy from the Atom   46
Energy Alternatives   48
Engines and Turbines   50

EXPLORING ELECTRICITY
Static Electricity   52
Current Electricity   54
Messages by Wire   56
Messages on the Air   58
Electrons and Chips   60

EXPLORING LIGHT
What is Light?   62
Seeing Near and Far   64
Colors and Waves   66
Laser Light   68

EXPLORING SOUND
Varieties of Sound   70
Sound Recording   72
Silent Sound   74

Index   76
Acknowledgements   77

MATTER AND THE UNIVERSE

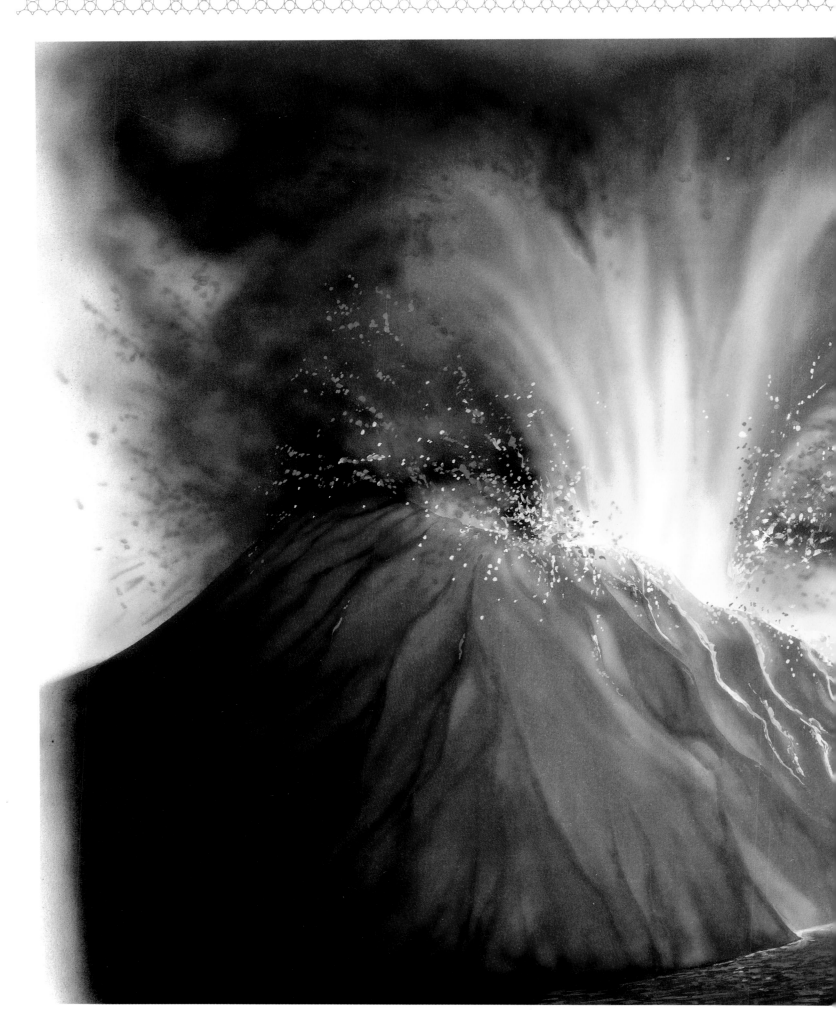

All the things around us – the ground, rocks, wood, metal, flesh, water, oil, and air – are different forms of what we call matter. The Earth, and all it contains, is an island of matter floating in space. The Sun and the stars are other, much larger, cosmic islands of matter, making up the bulk of the Universe.

On the Earth, matter appears in three main guises, or states. It is either a solid, like rock; a liquid, like water; or a gas, like air. However, matter does not always remain in the same state.

If you heat water in a pan for long enough, it boils away. The liquid water turns into a gas (water vapor) and disappears into the air. And if you leave the pan of water outside on a cold winter's night, you will find in the morning that it has turned to ice: a solid.

So, by changing the temperature, you have changed the state of water. And you can do the same for most other substances, provided you heat, or cool, them sufficiently. Even rock will become molten if heated enough (as happens in a volcanic eruption); and air will become liquid if cooled.

We explain how such a change takes place by what is called the kinetic theory of matter. Matter is made up of very tiny particles (the pieces into which matter breaks). In a solid, these particles are relatively close together and attract one another strongly. This attraction holds them in place. As the temperature rises, the particles gain more energy and try to break free. At a certain temperature (melting point), they can start to move around, as a liquid. As the temperature rises still further, they gain more energy still, until, at boiling point, they can escape from their neighbors completely, turning into a gas.

Left: Fountains of ash and choking fumes shoot into the air when a volcano erupts. Rivers of red-hot molten rock cascade down the slopes. The rock, usually solid, has been heated to melting point deep inside the Earth's crust. On the surface, it will eventually cool and turn into solid rock again. Rock formed in this way is called igneous ("fire-formed") rock.

CRYSTALS

Right: An uncut diamond. It does not look very attractive in this state. But when it is expertly cut, it will become an exquisite gem, sparkling with rainbow colors.

Below: Crystals of the mineral calcite, the same chemical compound as chalk. The shape of the crystals give it the name dog-tooth calcite.

Above: Pencil-like quartz crystals: The most common mineral in the Earth's crust, it is a form of silica, a compound of silicon and oxygen.

Left: Sulfur crystals, which are usually found naturally around vents (openings) leading from volcanic craters. Sulfur is not a chemical compound, but a chemical element.

Right: Metallic-looking crystals of galena, the compound lead sulfide. Like many other minerals, galena forms crystals which have a cubic shape.

If you look through a magnifying glass at the sugar in a sugar bowl, you will see that it is made up of thousands of tiny transparent cubes. These cubes formed naturally, when a solution of sugar was heated to drive away the water. They are sugar crystals.

A great many other solids also form well-shaped crystals. The crystals of different chemicals which join together are called minerals. They make up the rocks in the Earth's crust. In granite, for example, the colored pieces you see are different kinds of mineral crystals; usually black mica, pink feldspar, and glassy quartz. Even metals are made up of crystals. You will not usually see them because they tend to be all squashed together.

In rocks, too, the crystals are usually squashed together and can be any shape. But in places where there are cavities, mineral crystals can sometimes grow unhindered, forming exquisite shapes. Quartz is

GRAPHITE

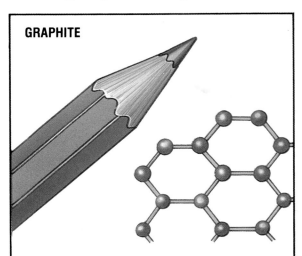

Pencil leads are made of a mixture of clay and graphite, one mineral form of carbon. Carbon is one of the softest of substances because of its flaky structure. Its atoms are arranged in flat sheets.

commonly found in these cavities, or geodes, as transparent pencil-thin columns. Calcite (crystalline chalk) forms a similar shape, called dog's-tooth calcite. The mineral fluorite forms transparent cubic crystals, tinged purple; galena, a lead mineral, forms metallic-looking cubes. Other minerals form needle-like, hexagonal (six-sided), octagonal (eight-sided), or pyramid-shaped crystals.

There are many crystal shapes, but each mineral forms only one shape, no matter where it is found. The shape is a reflection of the way the particles in the mineral are arranged at the atomic level (see page 20).

The most valuable crystals are not prized for their natural shape, but for the shape into which they can be cut. They are the precious stones, or gems. They include red ruby, blue sapphire, green emerald, and transparent diamond. All these crystals are hard, rare, and sparkle brilliantly when cut.

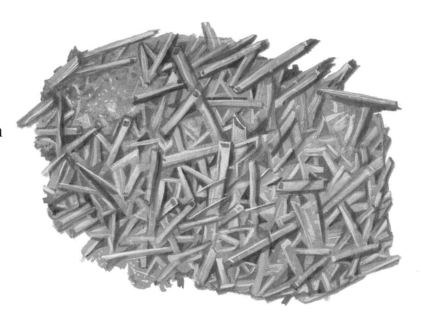

Below: Colorful crystals of crocoite, a lead chromate mineral. Many compounds of chromium with other elements are brightly colored.

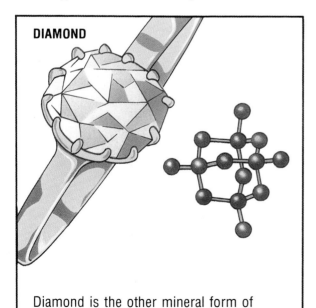

DIAMOND

Diamond is the other mineral form of carbon. Unlike graphite, it is very hard. In fact, it is the hardest substance there is. The reason is that its atoms are arranged in a rigid structure, which is difficult to break.

Left: The surface of galvanized steel, showing a mosaic of feathery crystals of zinc metal. Unlike ordinary steel, zinc does not rust or corrode. So when it is coated on steel, it makes the steel rustproof, too. The process of coating steel with zinc is called galvanizing.

Right: The mineral crystals show up clearly in this piece of granite from Cumbria, in northern England. The white and pink ones are different kinds of felspar; the black ones are biotite; and the glassy ones quartz. Granite is an igneous rock that formed when molten rock cooled slowly underground. There was time for quite large crystals to grow.

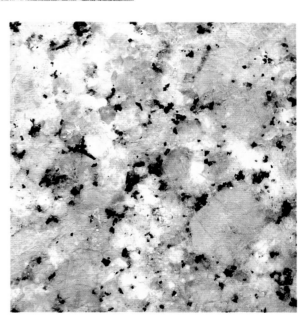

LIQUIDS

Below: A pond skater walks across the surface of a pond. Notice how its feet dent the "skin" on the water. The pond skater feeds on flies and other insects that get trapped by the skin.

Above: Torrents of water cascade over Niagara Falls, one of the natural wonders of the world, on the border between Canada and the U.S. Water is the most common liquid there is. The oceans that cover more than two-thirds of our world contain more than 832.8 million cubic miles of water. The water is not pure, but contains large amounts of salt and other dissolved substances.

Compared with crystals and other solids, liquids have very different physical properties. Solids are hard and have a definite size and shape. When heated or cooled, they do not expand or contract very much. Liquids, on the other hand, although they have a definite size, have no definite shape. They take the shape of whatever container they are in.

When heated or cooled, liquids expand or contract a lot. We use this property in simple thermometers. In the thermometer, the height of a thin column of liquid (usually mercury or alcohol) in a glass tube rises or falls as the temperature changes. This acts as a measure of that temperature.

Also, unlike solids, liquids can flow from place to place: they are fluids. When able, they flow under gravity from a high to a low level. When they are still, their surface is perfectly flat, or level. The ancient Egyptians realized this over 4,500 years ago when they began building the pyramids. To provide perfectly level foundations, they flooded channels on the sites, and marked where the water came.

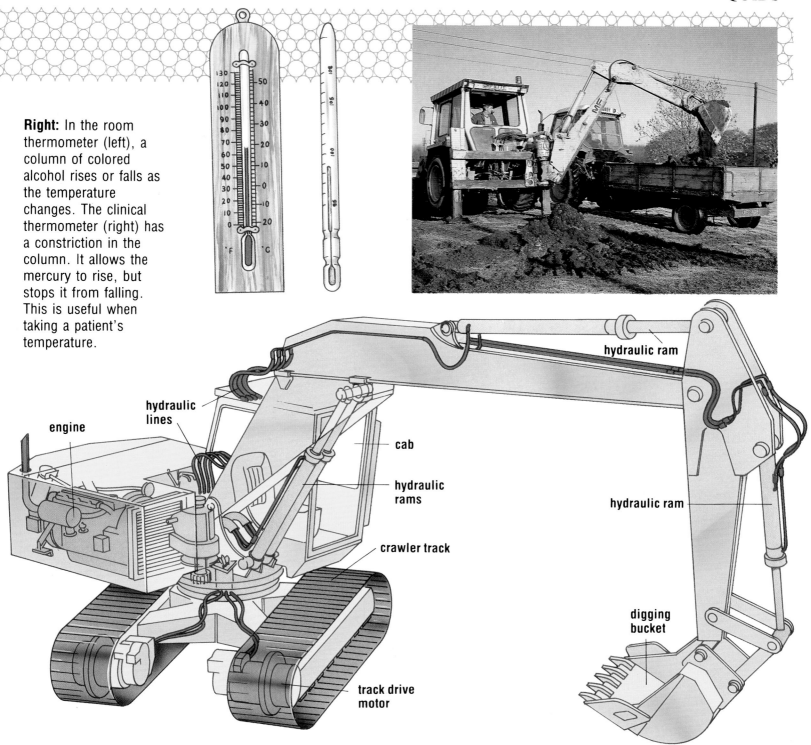

Right: In the room thermometer (left), a column of colored alcohol rises or falls as the temperature changes. The clinical thermometer (right) has a constriction in the column. It allows the mercury to rise, but stops it from falling. This is useful when taking a patient's temperature.

hydraulic ram

hydraulic lines

engine

cab

hydraulic rams

hydraulic ram

crawler track

digging bucket

track drive motor

Liquids have another interesting property – they have a kind of skin. If you look on the surface of a pond, you can often see tiny animals "walking" on the skin. The most common are known as skaters. If you look carefully, you can see where their hairy feet make little dents in the water's skin.

The skin happens because of unequal attraction between the water particles. The water particles exert a one-sided attraction on the particles on the surface, which produces a downward force that manifests itself as a skin. This force, called surface tension, is also responsible for the fact that drops of liquid and soap bubbles tend to form into spherical shapes.

The attraction of liquid particles for other materials gives rise to what is called capillarity. For example, water particles are more attracted to glass than to one another, so water will rise at the edges of a glass. In a narrow glass tube, the force of capillary attraction makes the water rise a long way. You can also see this effect quite clearly when you place a transparent straw in a soft drink.

Top: A JCB hydraulic excavator.
Above: This powerful excavator uses liquid, or hydraulic, pressure to transmit power. The liquid is pressurized by a pump driven by the engine. It is channeled through pipes to hydraulic rams. Inside the rams, the liquid pushes against pistons and forces them along cylinders. Rods attached to the pistons push and pull the arm and digging bucket.

LOOKING AT GASES

When you hold an empty glass, it is not really empty. If you put it upside-down under water, and then turn it the right way up, bubbles come out of it. The glass was in fact full – of air. Air is the most familiar example of the third state of matter – gas.

Air is all around us. We can't see it, smell it, or taste it. But we can feel it, when we breathe in and when we feel the wind blow; and we can see the effect it has on trees, water, flags on their poles, and so on. Gases are for the most part invisible because, compared with solids and liquids, their particles are very far apart.

A few gases are easier to detect than air. Hydrogen sulfide cannot be missed, because it has the revolting smell of rotten eggs! Sulfur dioxide has an unpleasant acrid smell. Chlorine has a distinct smell, too, familiar to us as the smell of swimming pools. The chlorine is used to disinfect the water, guarding against infection.

Gases vary widely in the ratio of their weight to the amount of space they occupy. This is known as their density. The lightest gas by far is hydrogen; helium, the next lightest, is four times heavier. But both are much lighter than air. That is why they have been used to provide the lift in gas balloons. Helium is almost always used today, because it does not burn. Hydrogen, on the other hand, forms an explosive mixture with air and burns readily.

Air is not a pure gas like hydrogen, but a mixture of gases. It is mainly nitrogen (78 percent), oxygen (21 percent) and argon (about 1 percent). There are also traces of many other gases,

Left: Hot-air balloons work on the simple principle that hot air rises. When you heat up air, or any other gas, it expands. As it expands, it becomes lighter than the air around it.

including carbon dioxide and water vapor. Oxygen is the gas that all animals and plants must breathe to live. Carbon dioxide is the gas living things give off when they breathe out. It is also notorious as the worst "greenhouse" gas. Its build-up is turning the atmosphere into a kind of greenhouse, which is trapping the Sun's heat and leading to a warmer climate. This could spell disaster for the long-term future of the human race.

Above: Inflating a hot-air balloon. The flames from a gas burner are directed into the open end of the fabric bag of the balloon. They heat up the air inside, which expands to fill the bag. Heating the air more makes it hotter and lighter, producing an upward force that makes the balloon lift up into the air.

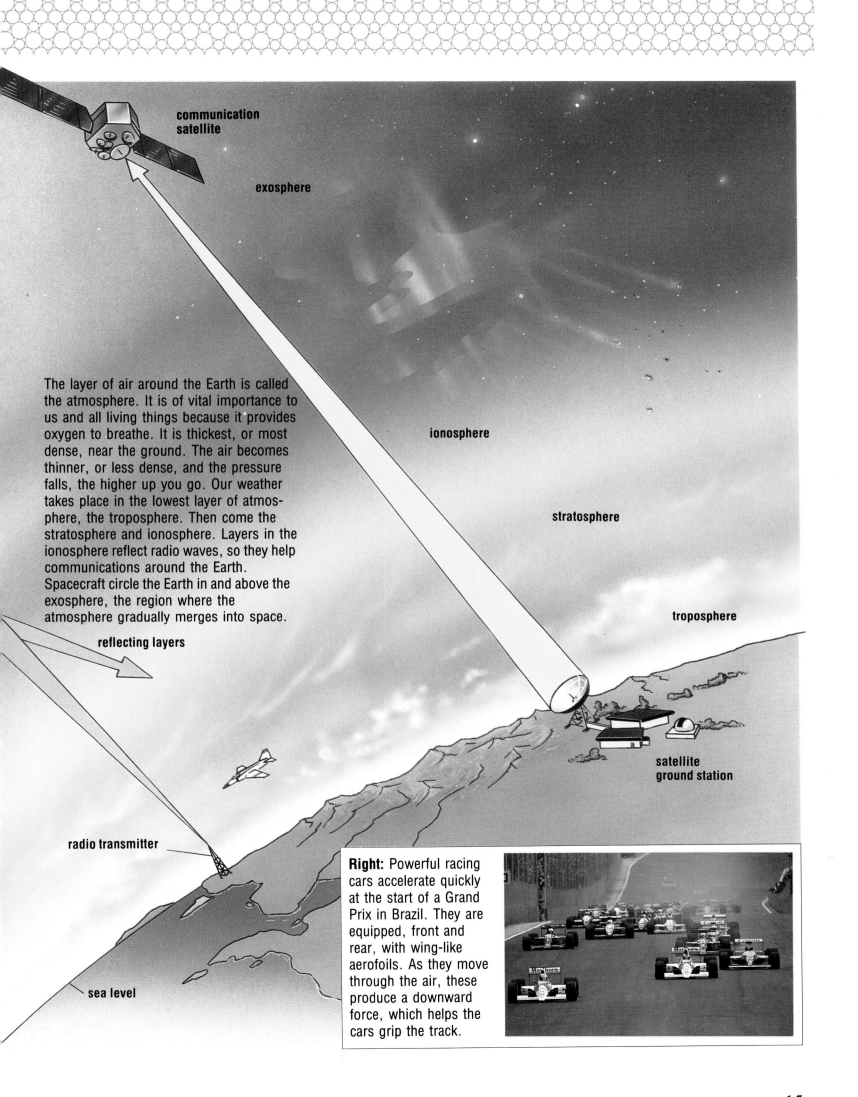

communication satellite

exosphere

ionosphere

stratosphere

troposphere

The layer of air around the Earth is called the atmosphere. It is of vital importance to us and all living things because it provides oxygen to breathe. It is thickest, or most dense, near the ground. The air becomes thinner, or less dense, and the pressure falls, the higher up you go. Our weather takes place in the lowest layer of atmosphere, the troposphere. Then come the stratosphere and ionosphere. Layers in the ionosphere reflect radio waves, so they help communications around the Earth. Spacecraft circle the Earth in and above the exosphere, the region where the atmosphere gradually merges into space.

reflecting layers

satellite ground station

radio transmitter

sea level

Right: Powerful racing cars accelerate quickly at the start of a Grand Prix in Brazil. They are equipped, front and rear, with wing-like aerofoils. As they move through the air, these produce a downward force, which helps the cars grip the track.

REACTION AND CHANGE

Above, main picture: Huge chunks of ice often break off the great ice sheets that exist in polar regions. They float away in the water as icebergs.
Inset, top: The rusting hull of a ship in Limassol harbor, Cyprus. Rusting is a chemical change brought about by the oxygen in the air.
Inset, bottom: This piece of stonework at Lincoln Cathedral has suffered chemical attack from acid rain.

Our world is in a perpetual state of change. For example, on cold nights, water on puddles freezes into ice. Next day, the Sun melts the ice back into water and evaporates the water into the air. High in the sky, the water vapor cools and condenses into water droplets, which gather into clouds. When the droplets get big enough, they fall from the sky as rain (or snow). Eventually, the rain or snow forms puddles on the ground, and the whole process begins again.

This process is never-ending and is known as "the water cycle." The transition of water from solid ice to liquid water, to gaseous vapor, and back again, is an example of a physical change that takes place in our world. When swift-flowing waters cut their way through the ground to create a river channel, or waves batter and wear away the cliffs, these are also physical changes. They are forms of erosion.

A host of chemical changes also take place in our world. If you leave a new steel nail outside for a few days in damp weather, it will lose its shine and acquire a reddish film, as it is slowly being eaten

away. Because of moisture in the air, the oxygen attacks, or combines with, the iron in the steel to form a new substance, forming the compound iron oxide, known better as rust (a process known as oxidation).

Burning is another oxidation reaction, which takes place much faster and gives out energy as heat and light. When coal burns, for example, the carbon it contains combines with the oxygen in the air to give the compound known as carbon dioxide.

The action of acids also demonstrates a common type of reaction. When carbon dioxide dissolves in river water, it forms a weak acid. When this flows over the rocks, it attacks their minerals and causes them to dissolve. This is a form of chemical erosion.

Factory chimneys spew out sulfur dioxide gas, which combines with moisture in the air to form sulfuric acid. This falls to the ground in the rain, attacking building stone, statues, and trees, and acidifying lakes so that they may no longer be able to support life. Acid rain is a serious pollution problem.

Inset, top: Fireworks contain a variety of chemicals that give off colored gases as they burn. The chemicals include compounds of strontium (crimson) and barium (pale green). Rockets are propelled by gunpowder, a mixture of carbon, sulfur, and saltpeter (potassium chlorate).
Inset, bottom: Chemicals in the body of a glowworm give off light when they react together.

BUILDING BLOCKS OF MATTER

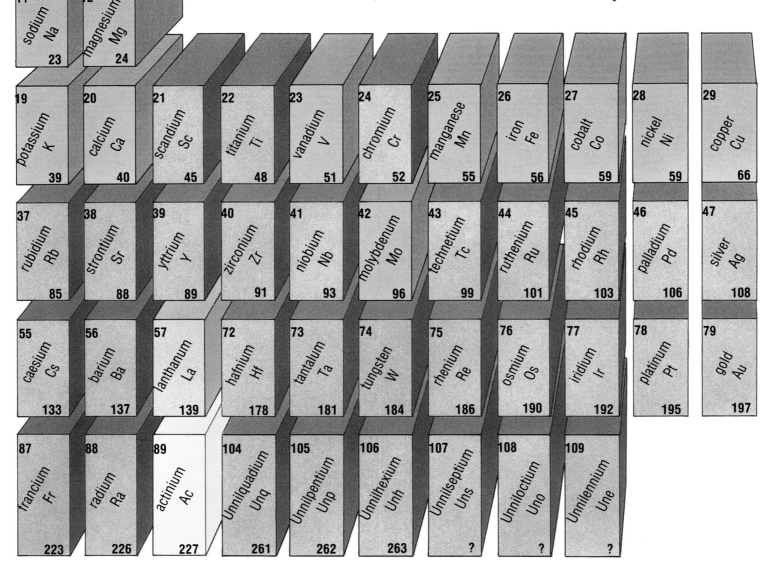

Look around you. How many different kinds of materials can you see? The chances are that you can see dozens, including perhaps wood, paper, steel, plastic, sugar, cotton, clay, water, steel, aluminum, and glass. In our world there are hundreds of thousands of different materials, each with different properties.

The material world, then, seems impossibly complex, but, in fact, it is not. If you have the right chemical tools, you can show that all materials are made up of about 90 basic substances. We call them the chemical elements. They are the building blocks of matter. Elements join together with one another chemically , in a variety of different ways, to form a host of chemical compounds. And that is

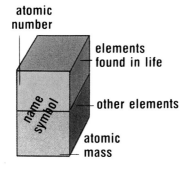

atomic number

elements found in life

other elements

name symbol

atomic mass

In the 1860s, a Russian scientist Dmitri Mendeleyev produced an arrangement of chemical elements known as the Periodic Table. A modern version of this Table is illustrated here, in which the elements appear in order of their atomic number (the number of protons in their nucleus). Elements with similar properties fall into the same vertical column, or Group. Elements in each horizontal row, or Period, show a gradual change in properties. The elements appear in these positions because of the structure of their atoms. The lanthanides (atomic numbers 57-71) and actinides (89-103) are two series of elements so similar that they appear as single blocks in the Table (see list opposite).

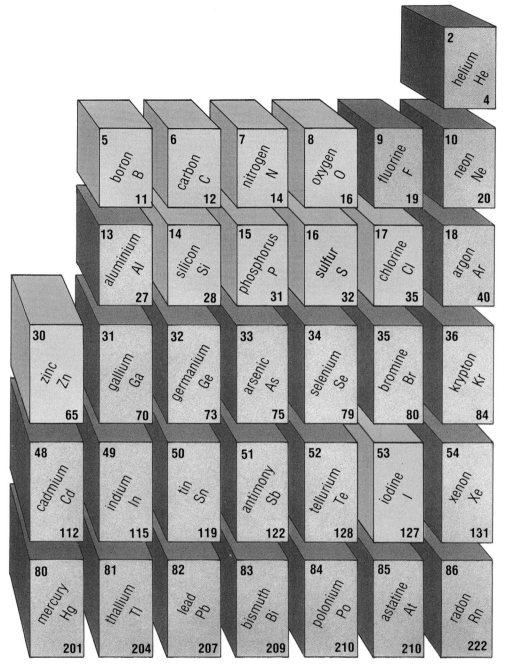

NAME	ATOMIC		
	NO.	SYMBOL	MASS
cerium	58	Ce	140
praseodymium	59	Pr	141
neodymium	60	Nd	144
promethium	61	Pm	145
samarium	62	Sm	150
europium	63	Eu	152
gadolinium	64	Gd	157
terbium	65	Tb	159
dysprosium	66	Dy	163
holmium	67	Ho	165
erbium	68	Er	167
thulium	69	Tm	169
ytterbium	70	Yb	173
lutetium	71	Lu	175
thorium	90	Th	232
protactinium	91	Pa	231
uranium	92	U	238
neptunium	93	Np*	237
plutonium	94	Pu*	242
americium	95	Am*	243
curium	96	Cm*	245
berkelium	97	Bk*	249
californium	98	Cf*	249
einsteinium	99	Es*	251
fermium	100	Fm*	253
mendelevium	101	Md*	256
nobelium	102	No*	253
lawrencium	103	Lw*	257

*man-made elements

what most of the materials we use are – chemical compounds.

Some are simple compounds, consisting of a few elements. Water, for example, is a simple compound. By passing electricity through it from a battery, you can split it up into two gases, hydrogen and oxygen. But no matter what you do chemically to these two gases, you cannot break them down further. They are basic building blocks – chemical elements.

Some of our everyday materials are not compounds, but elements themselves. The metals gold, silver, and copper are elements. So is carbon. We call them native elements, because they can be found as elements in nature.

All the chemical elements have different chemical and physical properties. Scientists have over the years investigated these properties in minute detail and have grouped the elements into this Table.

Above: This table lists the elements in the lanthanide and actinide series of elements, which appear in single blocks of the Periodic Table because of their similar chemical properties. All the actinides are radioactive; those from neptunium onwards are man-made.

INSIDE ATOMS

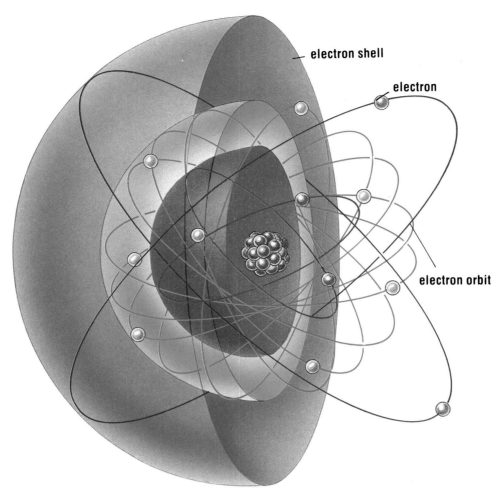

electron shell

electron

electron orbit

Above: A simple way of representing an atom. At its center is a nucleus, made up of two kinds of particles: protons and neutrons. Smaller particles, or electrons, orbit around the nucleus. Sets of electrons circle at different distances from the nucleus, in so-called shells. Here the shells are colored red, green, and blue. Two electrons circle in the inner shell, eight in the middle shell, and two in the outer shell. The chemical properties of an element largely depend on how the electrons in the outer shell link up with those of other elements.

The aluminum we use to make saucepans is a chemical element, or basic building block. If you could keep cutting it up smaller and smaller, you would eventually come to the smallest bit of aluminum that can exist. This is the aluminum atom.

Similarly, if you could keep cutting up the other chemical elements, you would eventually come to an atom. Each element has a different atom: its properties depend on the way its atoms behave and, in particular, the way they combine with other atoms.

The word "atom" comes from the Greek, and means "that which cannot be divided." However, scientists have proved this to be an inappropriate name for the tiniest bit of ordinary matter. With particle accelerators they can smash atoms to pieces, creating even tinier, subatomic, particles.

An atom is made up of three main particles: protons, neutrons, and electrons. Protons and neutrons are found at the center, or nucleus, of the atom, with electrons circling around them, like planets circling around the Sun in space. Protons have a positive (+) charge; neutrons have no charge, they are neutral; and electrons have a negative (−) electric charge. There are as many electrons as protons in each atom, so overall it is electrically neutral.

The thing that distinguishes one atom from another is the number of protons (or electrons) it has. This is called the atomic number. In the Periodic Table, the elements are arranged in order of their atomic numbers. The smallest particles of the compounds they form are called molecules.

Below: The circle in this picture of the countryside near Geneva, Switzerland, shows the site of a powerful particle accelerator, or atom-smasher.

TABLE OF SUBATOMIC PARTICLES

TYPE	Particle	Mass (Proton = 1)	Electric charge
BARYONS	Proton	1.00	+1
	Neutron	1.00	0
	Lambda	1.19	0
	Sigma	1.27	+1/0/−1
	Xi	1.40	0/−1
MESONS	Pions	0.15	+1/0/−1
	Kaons	0.53	+1/0
LEPTONS	Electron	0.0005	−1
	Muon	0.11	−1
	Tau	1.90	−1
	Neutrinos	0	0
QUARKS	Up	0.005	+2/3
	Charm	1.49	+2/3
	Top	?	+2/3
	Down	0.007	−1/3
	Strange	0.16	−1/3
	Bottom	1.92	−1/3
BOSONS	Photon	0	0
	Gluon	0	0
	Graviton?	0	0

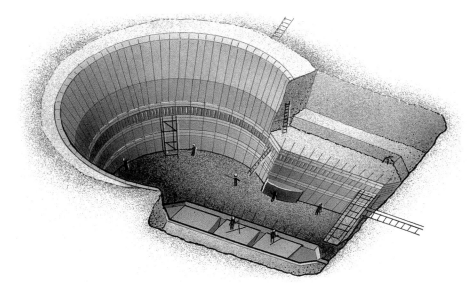

Above: An overhead view of a particle detector under construction at CERN, the European Center for Nuclear Research, near Geneva. The detector will record the tracks of particles created in collisions between high-speed particles accelerated in atom-smashers.
Right: Subatomic particles are pictured in this photograph, which was taken in a particle detector known as a bubble chamber. The particle tracks are artificially colored.

Above: Subatomic particles each have a different mass and a different electric charge. They also have a different direction of spin. Neutrinos are elusive particles with no mass or charge, only spin. Photons are the particles that make up beams of light. Gluons bind particles within the nucleus. No one has been able to detect gravitons, particles which might carry the property of gravity. Every particle is thought to be made up of quarks, probably the ultimate particles of matter.

RAW MATERIALS

Above: The Kennecott Copper Mine in Utah. The copper ore is extracted in a series of terraces, creating an ever-widening and ever-deepening pit. The ore is broken up by blasting with explosives and then loaded into trucks for removal.

Most of the materials we use in the modern world, are manufactured from other basic materials. Glass is made mainly from sand; metals from minerals; paper from wood; and plastics from crude oil. We call sand, minerals, wood, and oil, raw materials.

The sea, and even the air, also provide raw materials for manufacturing. For example, the salty sea provides us with table salt, the chemical sodium chloride. The air provides us with nitrogen gas, which manufacturers use to produce ammonia, from which fertilizers are made.

The bulk of our raw materials, however, are found in the ground as minerals and are extracted by mining. Mining is one of the largest-scale industrial activities of humankind. For example, the USA and USSR alone between them mine 300,000,000 tons of iron-bearing minerals each year.

Most iron, and aluminium, ores are extracted by surface, or opencast, mining. Various kinds of excavators, such as power shovels and bucket-wheel excavators, dig out the ore from the surface. This can create pits that scar the landscape.

Underground mining is necessary for most other ores, because the mineral layers, or veins, in the rocks lie hundreds of feet beneath the surface. It is carried out by drilling vertical shafts into the ground, and then driving horizontal tunnels from them to the veins. Explosives are generally used to break up the ore-bearing rock. Even today, underground mining is still very dangerous.

Crude oil, or petroleum, is extracted by drilling into the rocks. Holes are bored into rock strata (layers) where the oil has become trapped. Other minerals, including sulfur and salt, may also be extracted by borehole drilling.

Above: A bucket-wheel excavator at work in an opencast coal mine in Germany. The coal is a low-grade fuel known as lignite, or brown coal. It is soft enough to be scooped up by the rotating buckets.

METALS AND THEIR USES

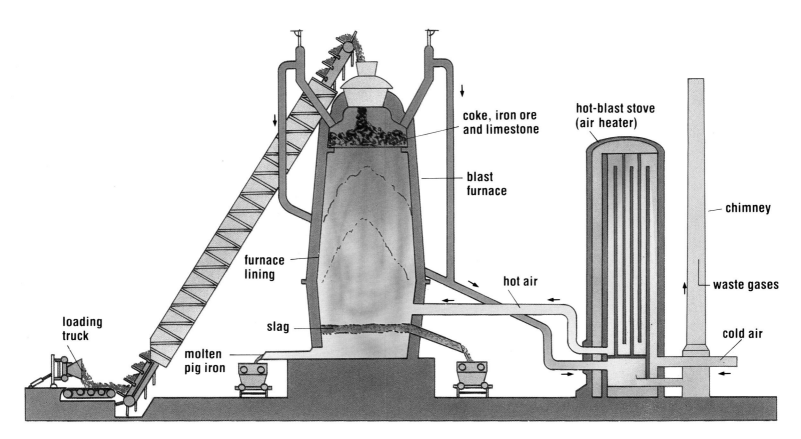

Above: Outline of a blast furnace. Iron ore, coke, and limestone are delivered into the top of the furnace. Inside, the coke burns fiercely, reducing the iron to a molten state, which trickles down into the base, or hearth, of the furnace. The limestone combines with impurities in the ore to form a molten slag, which collects on top of the iron. From time to time, the slag and ore are removed from the furnace through tapholes. The iron, still impure, is known as pig iron, because it was once cast in molds called pigs. Now it is usually transferred in a molten state to the steelmaking furnaces.

Most minerals need to be processed in some way before they become useful. For example, the minerals that contain metals – the ores – must usually be heated at a high temperature in a furnace. This process, called smelting, was discovered over 5,000 years ago. Its discovery marked the beginning of an "Age of Metals." Humans emerged at last from the Stone Age which had lasted for some two million years. For at least 2,000 years, iron has been the dominant metal. For a little longer than 150 years, we have used it mainly in the form of steel.

Iron ore is smelted into iron in a blast furnace, so called because hot air is blasted through it. Coke (a form of carbon) and limestone are charged into the furnace with the ore. The coke acts as fuel for the furnace. It also attacks the iron ore, which is iron oxide. It combines with the oxygen to form carbon monoxide gas, which

escapes from the furnace. Iron metal is left, in a molten state because of the high temperature (up to 2700°F or 1500°C). Earthy impurities in the ore combine with the limestone to form a molten mixture called slag.

Periodically, the furnace is tapped, and the slag and metal run off. Further purifying takes place in other furnaces, most widely in the basic-oxygen converter. In this furnace, a jet of pure oxygen is blasted into the molten iron. The impurities burn off, but a little carbon is allowed to remain, because iron plus a little carbon equals steel.

Steel is an iron alloy, a mixture of iron with another chemical element. In fact, we use most metals in the form of alloys (mixtures of different metals). Our coins are made of alloys: bronze (copper and tin) for our "copper" coins, and cupronickel (copper and nickel) for our "silver" coins.

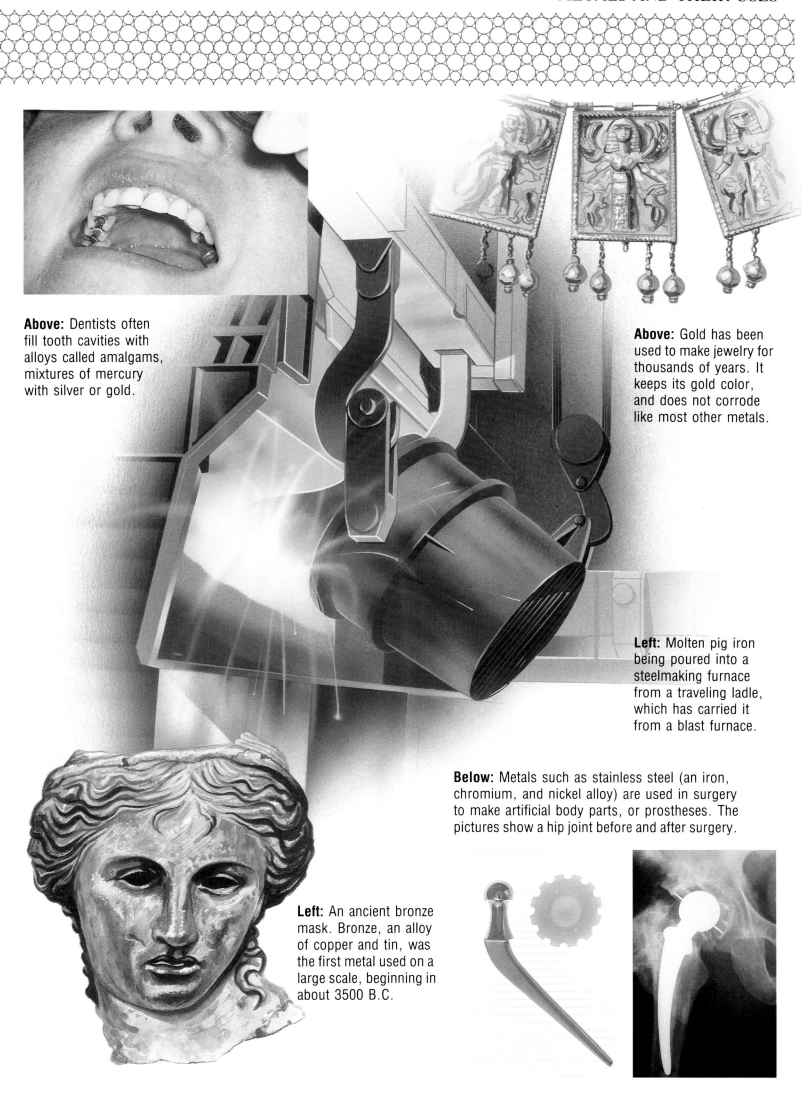

Above: Dentists often fill tooth cavities with alloys called amalgams, mixtures of mercury with silver or gold.

Above: Gold has been used to make jewelry for thousands of years. It keeps its gold color, and does not corrode like most other metals.

Left: Molten pig iron being poured into a steelmaking furnace from a traveling ladle, which has carried it from a blast furnace.

Below: Metals such as stainless steel (an iron, chromium, and nickel alloy) are used in surgery to make artificial body parts, or prostheses. The pictures show a hip joint before and after surgery.

Left: An ancient bronze mask. Bronze, an alloy of copper and tin, was the first metal used on a large scale, beginning in about 3500 B.C.

25

OIL DISTILLATION

Above: The oil refinery at Milford Haven in South Wales is one of the largest in Europe. It processes crude oil transported there by tankers from as far away as the Persian Gulf. A large refinery occupies much the same area as a small town, although only a handful of people are needed to operate it, because all the processes are highly automated.

Crude oil, or petroleum, has become one of the foremost raw materials for the organic chemical industry. Organic chemistry deals with the chemistry of carbon compounds. All living organisms are made up of such compounds, which is why their study is called "organic" chemistry.

Crude oil is a mixture of hundreds of organic compounds, mainly hydrocarbons, that is, compounds of hydrogen and carbon. Before it becomes useful, either as a fuel (see page 45) or as a chemical raw material, these

hydrocarbons must be separated out. This is done by means of distillation, a process that separates liquids according to their boiling point. The hydrocarbons in crude oil all have different boiling points.

The distillation of the oil takes place at a refinery. It is first vaporized in a furnace and then introduced into a tall unit called a fractionating column. At different levels up the column, there are a number of perforated trays. These are maintained at different temperatures, which decrease going upward.

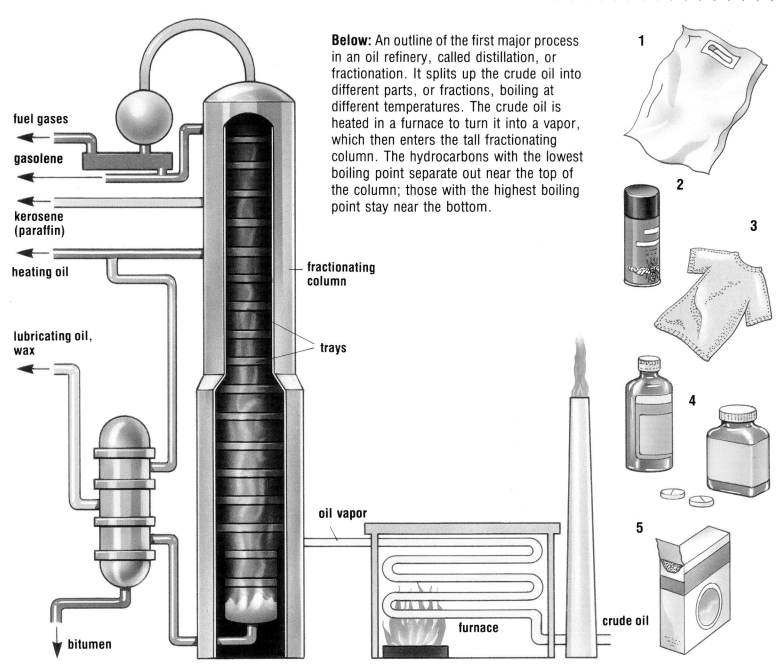

fuel gases

gasolene

kerosene
(paraffin)

heating oil

lubricating oil,
wax

bitumen

fractionating
column

trays

oil vapor

furnace

crude oil

Below: An outline of the first major process in an oil refinery, called distillation, or fractionation. It splits up the crude oil into different parts, or fractions, boiling at different temperatures. The crude oil is heated in a furnace to turn it into a vapor, which then enters the tall fractionating column. The hydrocarbons with the lowest boiling point separate out near the top of the column; those with the highest boiling point stay near the bottom.

1

2

3

4

5

The oil vapor passes up through the trays, and each hydrocarbon in it condenses, or turns back into liquid, in the tray where the temperature is just below its boiling point. In practice, a number of hydrocarbons with a variety of boiling points condense in each tray, making up what is called a fraction.

The lighter fractions near the top of the column, are most useful as fuels, including gasoline, kerosene and diesel oil. The heavier fractions at lower levels are by themselves less useful. But in further

processing, they can be broken down, or cracked, into lighter compounds for use in gasoline.

Both cracking and straight distillation yield a lot of light gases. They are converted into more useful products by another refinery process, called polymerization. This is the opposite of cracking. It combines small, light molecules into large, heavier ones, again producing more valuable products, such as gas. And, at each stage of processing, the refinery produces a host of organic chemicals, often called petrochemicals.

Above: An enormous variety of products is made from chemicals obtained by refining crude oil. The products include (**1**) Plastics, from which, for example, shopping bags can be made. (**2**) Insecticides, used to kill insect pests. (**3**) Synthetic (artificial) fibers, used in textiles. (**4**) Pharmaceuticals, or drugs, used to treat diseases. (**5**) Detergents, compounds that cleanse fabrics much better than soap.

VERSATILE PLASTICS

One of the main gases produced during cracking at a refinery is ethylene (or ethene). It has small molecules, made up of two carbon (C) atoms, to each of which are joined two hydrogen (H) atoms. We can write it symbolically as $H_2C=CH_2$. When this gas is polymerized at the refinery, its small molecules start joining together, until they form a very long chain.

The resulting long-chain molecule is called a polymer (meaning "many parts"). We know it as one of our most common plastics, polyethylene. The other materials we call plastics are made by polymerization in a similar way, and also consist of long-chain molecules with a "backbone" of carbon atoms. Among the chemical elements, only carbon can join up with itself to form molecules with long chains. Many compounds in living things have molecules consisting of long carbon chains.

All plastics can be shaped easily, by first heating them until they soften, and then molding them. Some, including polyethylene, soften again when reheated. These so-called thermoplastics also include nylon, PVC (polyvinyl chloride), and styrofoam.

Another class of plastics, however, will set rigid when first heat-molded into shape. Called thermosets, they include the original synthetic plastic, Bakelite. Bakelite, made from phenol and formaldehyde, was discovered in the United States by the Belgian-born chemist Leo Baekeland in 1909. It gave birth to what is now one of the most important industries in the world.

Chemists have created several hundred different kinds of plastics, which have found widespread use in the home, both outside and inside.

Above: Plastics abound in the kitchen. The floor is covered with decorative vinyl plastic, which is attractive to look at and easy to clean. The worktop surfaces are plastic laminates ("sandwiches") made of thermosetting plastic. They are heatproof and easy to wipe clean. Kitchen gadgets such as food mixers are molded in thermoplastics of various kinds, as are bowls, "squeeze" bottles, and baskets.

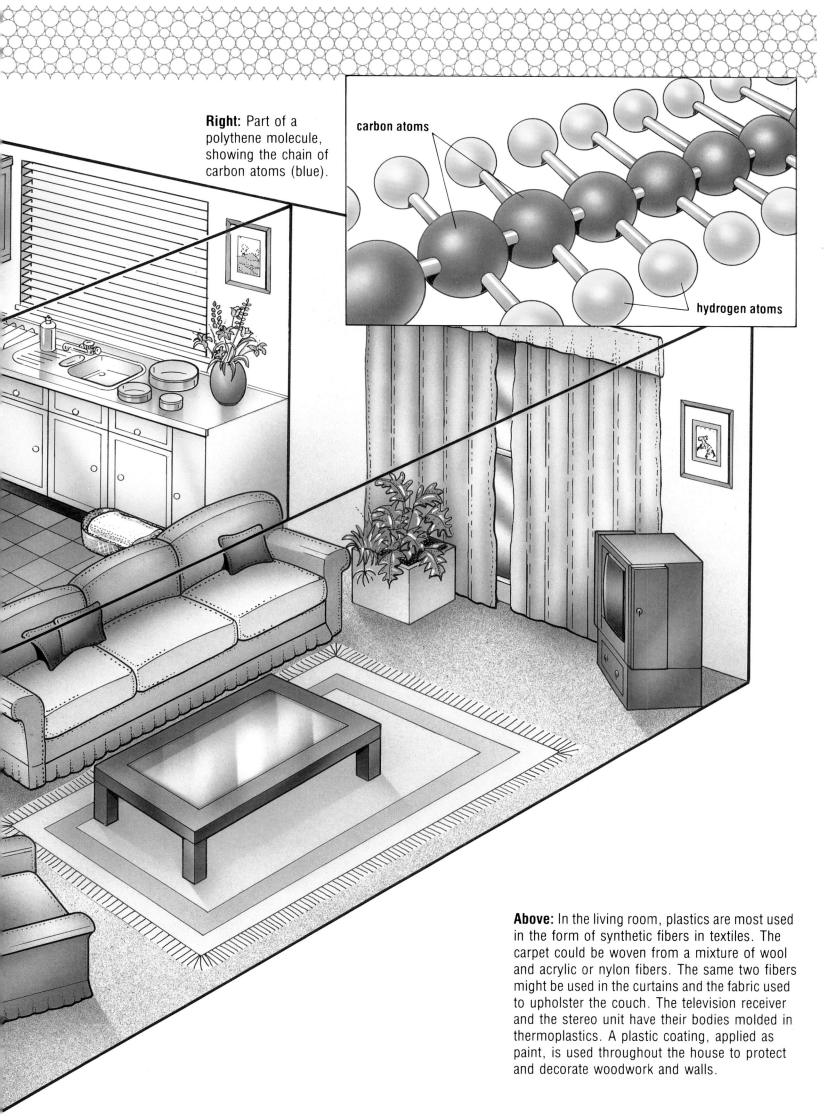

Right: Part of a polythene molecule, showing the chain of carbon atoms (blue).

carbon atoms

hydrogen atoms

Above: In the living room, plastics are most used in the form of synthetic fibers in textiles. The carpet could be woven from a mixture of wool and acrylic or nylon fibers. The same two fibers might be used in the curtains and the fabric used to upholster the couch. The television receiver and the stereo unit have their bodies molded in thermoplastics. A plastic coating, applied as paint, is used throughout the house to protect and decorate woodwork and walls.

PESTICIDES AND DRUGS

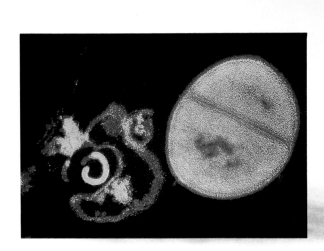

Above: This electron microscope picture shows in false color how an antibiotic has attacked and broken down a bacterium (left). **Above (main picture):** A crop-spraying plane at work. Spraying from the air is the most effective way of treating large areas of growing crops. Chemical spraying helps eliminate insect pests and fungus diseases, which could otherwise greatly reduce the yield of a crop.

The chemical industry, using both mineral raw materials and petroleum chemicals, produces a vast array of other products on which our modern civilization depends. They include the basic chemicals such as sulfuric acid, used in all kinds of industrial processing; synthetic fibers such as nylon; dyes and detergents; adhesives and antifreezes. But two of the most important products of the industry are pesticides and drugs, or pharmaceuticals.

More than anything else, the world has pesticides to thank for the great increase in food production which has taken place in the last few decades – the so-called "Green Revolution." The farmers' crops have always been at the mercy of insect pests and fungus diseases, and had their growth stunted by weeds. Nowadays, farmers can declare chemical warfare on these pests, with powerful insecticides, fungicides, and herbicides.

The insecticide DDT provided the first major breakthrough in insect control in the 1940s. But compounds such as this, which are hydrocarbons containing chlorine, are now banned in many countries because they harm not only the destructive insects, but friendly ones as well, such as bees and lady bugs. Also, they get into the food chain not only of other wildlife, but also of humans. Pesticides more friendly to the environment are now used in their place, many based upon organic phosphorus compounds.

The first major drug made by the chemical industry was aspirin, a

Above: Drugs are usually taken either orally (by the mouth) or by injection. For example, the vaccine against the paralysing disease poliomyelitis (polio) is usually given orally on a sugar cube, while that against tetanus is given by injection.

century ago. It is still the world's most commonly taken pain-killer, or analgesic. Thousands of other drugs are now on the market to treat the hundreds of diseases that afflict us. Many are man-made by chemical processing; others, the antibiotics, are produced by molds and other micro organisms.

The original antibiotic, penicillin, came into use in the 1940s, allowing treatment of many formerly deadly diseases, such as pneumonia. It was the beginning of what is now termed biotechnology, the manufacture of useful products from living organisms.

Biotechnologists are even able to improve on nature by altering the genes of micro-organisms to produce specific drugs. This is an example of a technique called genetic engineering.

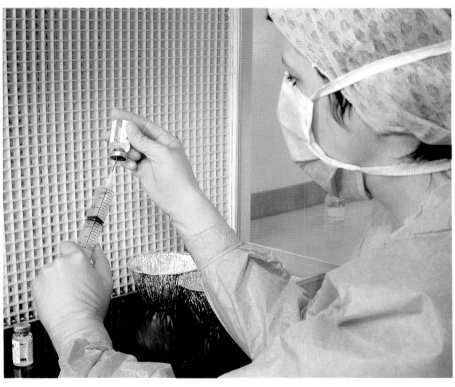

Above: A pharmacist in a hospital fills her hypodermic syringe with antibiotics from a sealed container. To give an injection, the thin hollow needle of the syringe is pushed under the skin, and the plunger is pressed down to force a measured quantity of drug into the body. This kind of injection is called subcutaneous. An intravenous injection is made into a vein.

MOVEMENT

Below: Jet planes are swift movers. Jets of hot gases rush backward out of their engines and force them forward by reaction.

When a soccer ball is resting on the ground, it will stay there unless you push it, pull it, or kick it. If you do kick it, it will keep on moving unless something stops it, such as a goalkeeper's hands. You could describe these two events as a scientific law: a body (ball) will remain at rest, or continue moving, unless it is acted upon by a force (push, pull, kick). The English scientist, Isaac Newton, summed up the action of forces on matter more than three centuries ago. It is called his First Law of Motion.

The tendency of a body to remain in the state it is, either stationary or moving, is called its inertia. In the real world, inertia can be dangerous. Imagine you are cycling, and a stick accidentally gets jammed in your front wheel. The wheel will stop dead; but, because of inertia, your body will continue moving forward and you could shoot over the handlebars.

Newton also realized that a force applied in one direction gives rise

Left and Below: On the ground, wheels make efficient transportation possible. Pedal power and engine power are needed to overcome friction with the ground.

to an equal force in the other. If you are on roller skates and push against a house, the house pushes back and pushes you away. When fuel burns in a jet engine, it produces hot gases that shoot out backwards. This force gives rise to a forward force, which propels the jet plane. Newton summed up this effect in his Third Law of Motion: to every action (force in one direction), there is an equal and opposite reaction (force in the opposite direction).

In our world, the force of the wind causes trees to bend; the force of gravity causes objects to fall; and so on. Also, there are hidden forces that oppose movement. A ball rolling across the ground will soon stop because a force is slowing it down, the force of friction. This force is set up by the ground rubbing against the ball. Friction exists between any two surfaces rubbing together. Even the air exerts friction, or drag, which slows down planes and cars.

Above: On "big dipper" roller coasters, the rapid circular movement of the passenger cars produces forces that make them defy the pull of gravity.

Above: A carnival merry-go-round goes round in circles. It is a big wheel, driven by sets of toothed gear wheels, which are turned by machines with spinning gear wheels and rotating shafts.

Above: The force of moving air, or the wind, bends trees and can cause damage to property in storms.

Left: A ball will stay where it is on the ground until someone applies a force to it by kicking.

FALLING THINGS

Left: A portrait of the English scientist Isaac Newton, who lived from 1642 to 1727. He pioneered many scientific and mathematical studies, discovering the laws both of motion and gravity.

Right: A pair of simple, equal-arm scales, used for comparing weights. An object of unknown weight is placed in one scale tray, and weights of known value are added to the other tray until the arm balances.

Right: The package of butter and feather-filled pillow both have the same mass, 1 lb. They have a different size, however, because they have a different density, or mass, per unit volume. The package of butter is smaller because it has a greater density.

Gravity keeps our feet firmly on the ground and gives us our sense of up and down. Down is the direction in which things will fall when they are dropped; up is in the opposite direction. We can temporarily overcome gravity by movement, say, by throwing something up into the air. But it doesn't go far before gravity pulls it back down to Earth. Birds and planes overcome gravity when they fly through the air. Their forward motion sets up a pressure difference between the top and bottom of their wings. This results in a force acting upward, called lift. When they are moving fast enough, the lift overcomes their weight — the pull of gravity on them — and they become airborne.

Above: Free-fall parachutists leap from airplanes and drop like stones before they operate their parachutes. They can reach a maximum speed called the terminal velocity, about 185 mph.

When an apple is ripe, it will fall from the tree. It falls because it experiences a force. We call this force gravity. It is a force that the Earth exerts on everything in, on, and around, it. Without gravity, nothing could stay on the Earth: no people, no animals, no oceans, no air. Indeed, there would be no Earth, nor any of the other heavenly bodies, because gravity is what holds the universe together.

Gravity is responsible for an object's weight. Weight is the downward force exerted on the object by gravity. This force varies with mass; that is, the amount of matter in the object. An object with more mass has more weight.

Mathematically, the weight of an object is expressed as the product of its mass and the acceleration due to gravity (g). This is the rate at which everything falls to the ground when dropped. Weight is expressed scientifically in units of force called newtons (N), when the mass is expressed in kilograms (kg).

On the Earth, g is about 10 m/s^2 (32 ft/sec^2) (that is, 10 metres (32 feet) per second per second). A bag of sugar with a mass of 1 kg (2.2 lbs) weighs about 10 N (70.4N) (mass × g). In everyday life, however, we talk about the bag *weighing* 1 kg (2.2 lbs), but that is not strictly accurate. In science, kg is a measurement of weight and mass, lbs only measure weight.

On the Moon, the weight of the bag of sugar would be much less than on the Earth. The Moon has only one-sixth of the Earth's gravitational pull, that is, its g is only 1/6 × 10 = 1.66. So the bag will weigh only 1.66 N. Its *mass* of 1 kg will remain the same.

BEATING GRAVITY

Left: Gravity is the most widespread force in the universe. It binds the stars into great galaxies, like this one in the heavenly constellation Andromeda. The Andromeda galaxy is one of the very few outer galaxies we are able to see with the naked eye.

Above: When you whirl a chestnut on a piece of string around your head, you can feel it pulling outward on the string. To keep it traveling in a circle, you have to pull in on the string. In a similar way, when a satellite is traveling in orbit around the Earth, it is trying to pull away, but the inward pull of gravity stops it. To escape completely from Earth's gravity, a spacecraft must travel at a speed of no less than about 25,000 mph.

Gravity exerts a powerful pull on everything on Earth, so how can we beat gravity, and launch spacecraft into space?

If you throw a ball up into the air, it rises so far and then gravity pulls it down. You could make the ball go much higher by fitting a rocket to it, but eventually gravity will again pull it down. However, this gives you the clue to beating gravity – speed. If you fitted a very powerful rocket to your ball, you could make it overcome gravity and enter space.

The minimum speed that you would have to give your rocket to get into space is no less than 17,400 mph, or nearly 30 times the speed of a jumbo jet! At this speed, you could make your ball circle round and round the Earth, in orbit. It would become a satellite, following in the path of Sputnik 1, the Russian satellite that launched the Space Age on October 4, 1957.

Today, it would join several hundred satellites in orbit, which circle at heights from a few hundred miles, to tens of thousands of miles. The most useful are communications satellites, which relay telephone and television signals, fax messages and computer

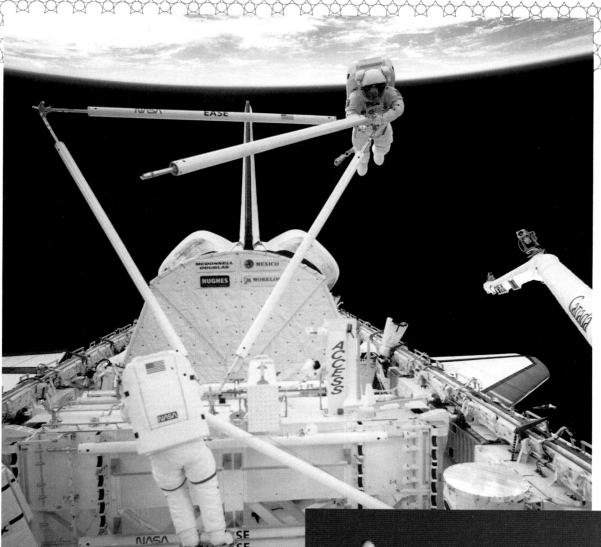

Right: The space shuttle orbiter "Atlantis" blasts off from the launch pad at the Kennedy Space Center. Its three main engines and the two massive booster rockets all fire together to produce a lift-off thrust of more than 6½ million lbs.

data. Weather and Earth-survey satellites are also important.

Most communications satellites, including the Intelsats, orbit 22,400 miles high over the equator, and take just 24 hours to circle the Earth once. During this time, however, the Earth also turns around once, so in effect the satellites appear fixed in the sky. Such an orbit is called a geostationary orbit. Some weather satellites, such as Europe's Meteosat, are also placed in such an orbit, where they can view weather conditions over a vast area of the Earth.

MAGNETISM

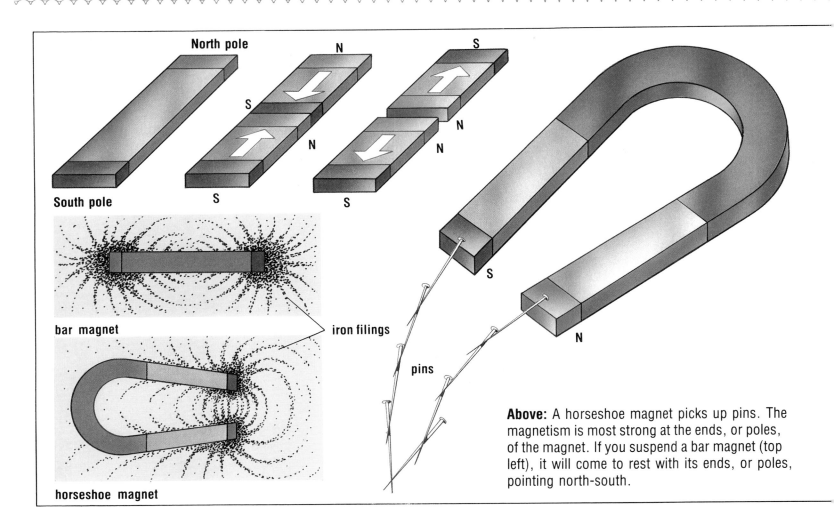

North pole

South pole

N

S

S

N

S

N

N

S

N

bar magnet

iron filings

horseshoe magnet

pins

S

N

Above: A horseshoe magnet picks up pins. The magnetism is most strong at the ends, or poles, of the magnet. If you suspend a bar magnet (top left), it will come to rest with its ends, or poles, pointing north-south.

Above: Iron filings arrange themselves into patterns when they are sprinkled around a bar magnet (top) and a horseshoe magnet (bottom). They form into curves because they follow the direction of the magnetic forces. The region around a magnet where magnetic forces act is called the magnetic field.

The pull of gravity is an invisible force that exists between lumps of matter. Magnetism is another property of matter that sets up invisible forces. This phenomenon is confined on Earth to lumps of iron, and a few closely related metals, such as cobalt and nickel.

Some iron minerals are natural magnets, including one called lodestone. The ancient Greeks and Chinese were familiar with the way bits of lodestone stuck to one another or pushed each other away – two basic properties of magnets.

The Chinese were probably the first to discover another fascinating property of magnets, that when they are suspended, they always come to rest pointing in the same direction. This led in the 13th century, to sailors using pieces of lodestone as a compass, which

helped them find their way across the seas. The word "lodestone" means "guiding stone."

When a magnet is suspended, it comes to rest pointing north-south. We call the ends of the magnet north and south poles, according to which direction they point.

From the way a suspended magnet and a compass behave, you would think that the Earth itself contained a huge magnet whose influence, or field, could be felt. However, there is no Earth magnet as such. Its magnetism is thought to be set up by electric currents flowing through its central core.

This explanation seems probable because magnetism and electricity are closely related. Study of the two is called electromagnetism. A magnetic field is always set up when an electric current flows in a wire.

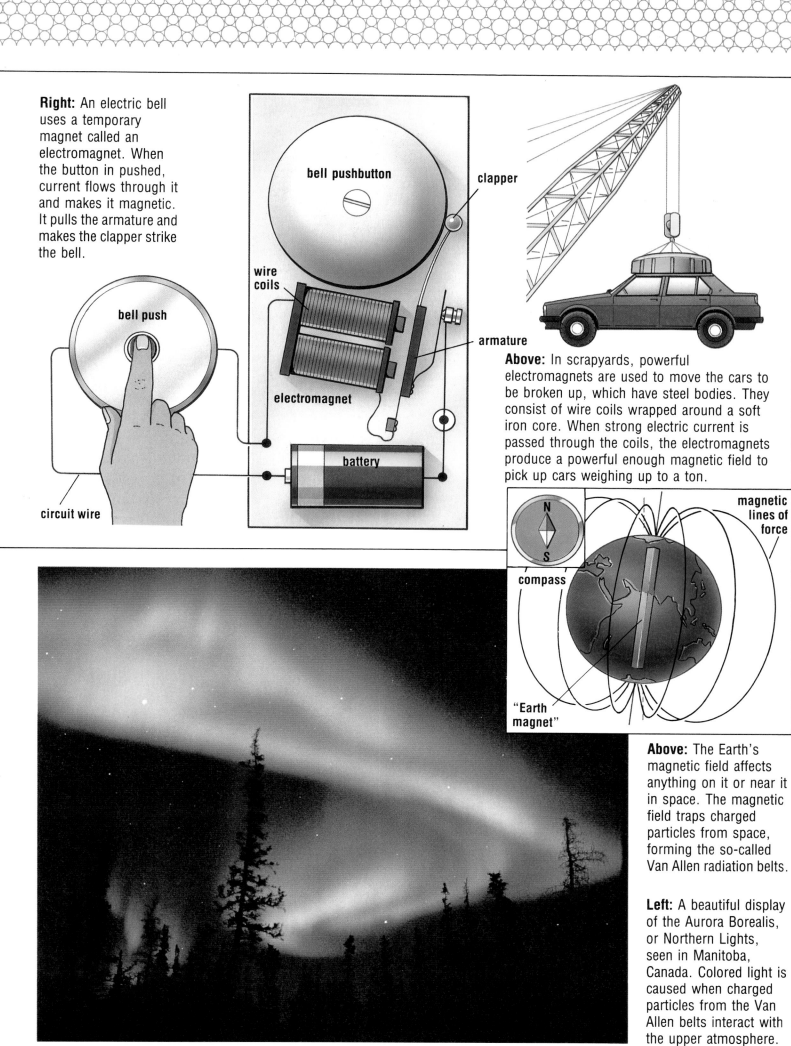

Right: An electric bell uses a temporary magnet called an electromagnet. When the button in pushed, current flows through it and makes it magnetic. It pulls the armature and makes the clapper strike the bell.

bell pushbutton

clapper

wire coils

bell push

electromagnet

armature

battery

circuit wire

Above: In scrapyards, powerful electromagnets are used to move the cars to be broken up, which have steel bodies. They consist of wire coils wrapped around a soft iron core. When strong electric current is passed through the coils, the electromagnets produce a powerful enough magnetic field to pick up cars weighing up to a ton.

magnetic lines of force

compass

"Earth magnet"

Above: The Earth's magnetic field affects anything on it or near it in space. The magnetic field traps charged particles from space, forming the so-called Van Allen radiation belts.

Left: A beautiful display of the Aurora Borealis, or Northern Lights, seen in Manitoba, Canada. Colored light is caused when charged particles from the Van Allen belts interact with the upper atmosphere.

SIMPLE MACHINES

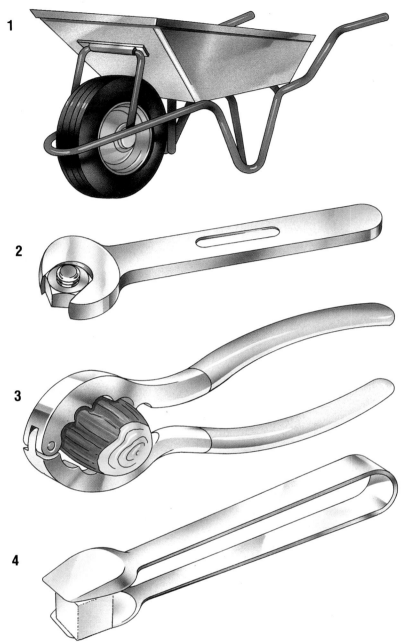

If one of our early ancestors wanted to move a big boulder, he or she would place the end of a stick beneath it and a smaller rock behind the stick, and then pull on the free end of the stick, using the small rock as a pivot. In this way, he could magnify his puny pulling power enough to dislodge the boulder.

Our boulder-shifting ancestor was using the stick as a simple machine. Machines can be described broadly as devices that apply forces to carry out useful work. The stick, as used above, is a simple machine we call a lever. It was able to magnify pulling power because of the way it was set up. It applied a small force (muscle power) over a long distance (movement of handle) to produce a large force (to lift the boulder) over a small distance. Other kinds of levers include seesaws, wheelbarrows, sugar tongs and pliers. They apply leverage forces

Some examples of different kinds of levers, which are among the simplest machines. Levers can be divided into different types, or classes, according to where the fulcrum (pivot) is and where the effort and load, or force produced, act. In the wheelbarrow (**1**), the fulcrum is at the wheel end, the load is in the middle, and the effort is applied at the handles. This is an example of a second-class lever. Nutcrackers (**3**) and bottle openers (**5**) are also second-class levers, with the fulcrum at one end and the load between it and the effort. Sugar tongs (**4**) and the human forearm are examples of third-class levers, with the effort between load and fulcrum. The brace-and-bit (**6**) is an example of the wheel and axle. Effort is applied to turn the handle in a large circle to make the bit turn a small circle. The wrench (**2**) is another example.

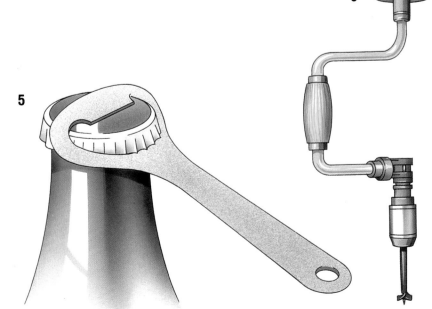

in different ways and have their pivots, or fulcrums, in different positions.

The wheel-and-axle is another simple machine; an example would be a hand-driven winch. The winding cable is wound around an axle with a small diameter, while the winding handle turns in a large diameter circle. Applying a small force around a large circle produces a magnified force on the cable, which moves around the axle in a small circle.

Even a sloping plank can be considered a simple machine, called an inclined plane. Pushing a heavy chest into the back of a truck up a sloping plank is much easier than lifting it vertically. Screws are another simple machine, whose thread is like a spiral inclined plane. A screw jack is a lifting machine that uses a small effort applied over a long distance by the winding handle, to lift a large load (like, a car) a short distance.

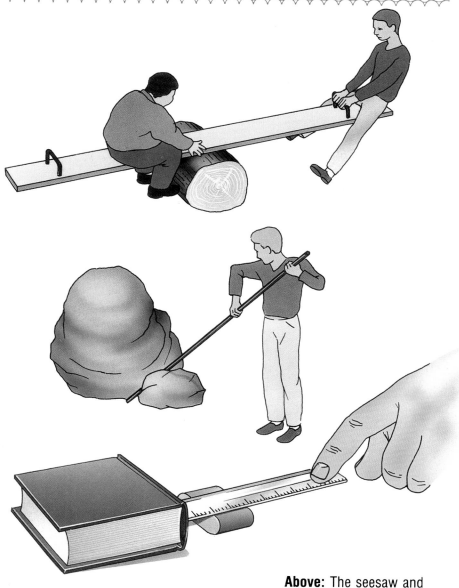

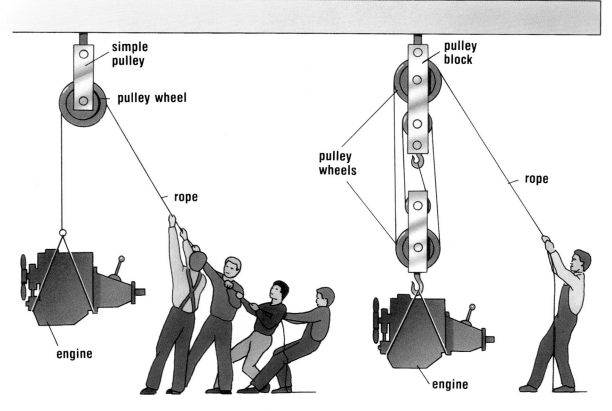

Above: The seesaw and crowbar are examples of first-class levers, in which the effort and load act on opposite sides of the fulcrum. On a seesaw, a light person sitting at the end on one side can balance a heavier person sitting on the opposite side closer to the fulcrum. Using a crowbar, you can shift heavy loads. Using a ruler as a lever, and an eraser as a fulcrum, you need little effort to lift a book.

Left: Using a set of pulleys, or a pulley block, allows you to produce a powerful lifting force with very little effort.

HOT AND COLD

Picnicking at the beach on a sunny day in summer. This is the season of the year in which the Sun climbs highest in the sky at noon and brings us the hottest weather. In winter, the Sun does not climb very high in the sky and does not bring us much heat.

Left: Hang gliders riding on air currents called thermals. They are upward currents caused by the air above the hot land itself becoming heated and rising.

It is nice on hot days to keep sandwiches, chocolate and iced cakes cool. These have been carried in a cool box lined with plastic foam. The air bubbles in the foam act as insulation to prevent the heat outside from flowing in.

All forces, and machines that apply forces, require the input of energy. The force of the wind requires the input of heat energy from the Sun. When we kick a ball, our muscles require the input of chemical energy from the body. We must supply a car engine with heat energy to drive its pistons. We must supply a vacuum cleaner with electrical energy to turn its motor.

Applying energy via forces in general gets things moving. When they move, they have what we call kinetic energy. But even a stationary object has energy. When you hold an orange in your hand, it has potential energy, the energy of position. If you let it go, it will start moving downward, accelerated by the force of gravity. It then acquires kinetic energy.

Of all the forms of energy, the most familiar is heat. It is a form our bodies can sense. We rely on the heat energy put out by the Sun for our existence. Without it, the Earth would be deathly cold and uninhabitable.

The Sun's heat travels through space, as a form of radiation. When heat from a hotplate heats up a skillet, the heat travels by conduction, or contact. The heat makes the molecules of the pan in contact with the hotplate vibrate more vigorously, and this vibration is passed from molecule to molecule. So the heat travels. The property of matter that we call temperature is a measure of the energy of vibration, or kinetic energy, of the molecules.

The hotplate also heats up the air immediately above it. This makes the molecules in the air travel faster, so the air expands. The expanding air is less dense than the cold air around it and so it rises. Air currents are set up, with hot air rising, and cold air moving in to take its place (convection).

Left: These wind surfers are catching the light winds we call sea breezes. Winds blow over the surface of the Earth mainly because of differences in temperature found from place to place.

Right: This girl's skin has become tanned brown. Invisible ultraviolet rays in sunlight cause tanning. But you should only sunbathe a little at a time; otherwise, the rays will burn you.

Ice cream is usually kept cold in a freezer. But the heat from the Sun, and a warm tongue, will soon melt it.

FOSSIL FUELS

direction
of air flow

Bottom: A miner checks the operation of a shearer, one type of coal-cutting machine. He works under the protection of the "walking props".

Below: In a coal mine, vertical shafts may extend several miles deep into the ground. There are usually two main shafts, used for elevators and ventilation, carrying fresh air into, and stale air out of, the mine. The coal is cut from a long face and carried along the tunnels by conveyors. Roof supports called "walking props" protect the cutting machines.

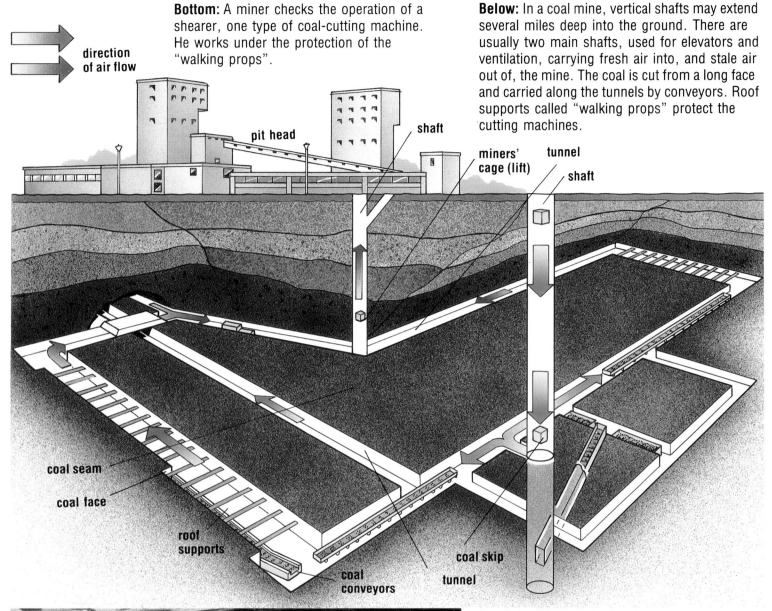

pit head

shaft

miners'
cage (lift)

tunnel

shaft

coal seam

coal face

roof
supports

coal skip

coal
conveyors

tunnel

Hundreds of millions of years ago, the Earth was covered in warm, shallow seas, which teemed with minute plant and animal life. There were also vast swamps, in which great forests of gigantic trees and ferns grew.

Eventually, the marine life forms died and fell to the bottom of the seas. They were covered with sediment and began to decay as rocks. They were buried under yet more sediment, then heat and pressure in the rocks turned them into oil and gas. The oil and gas gradually filtered through the rocks until they became trapped. Today,

Right: In some oil-rich regions of the world, oil lies in rocks under the sea. The picture shows an oil-production rig off the island of Sumatra, in Indonesia. There are also rich offshore oilfields in the Gulf of Mexico, Venezuela and the North Sea. In the North Sea, deposits are found in rocks several thousand yards under the seabed, which may itself be some 650 ft below the surface. Drilling for oil at such depths and under such circumstances is difficult, and advanced technologies have been pioneered to do it. The oil is extracted using production platforms anchored to the seabed.

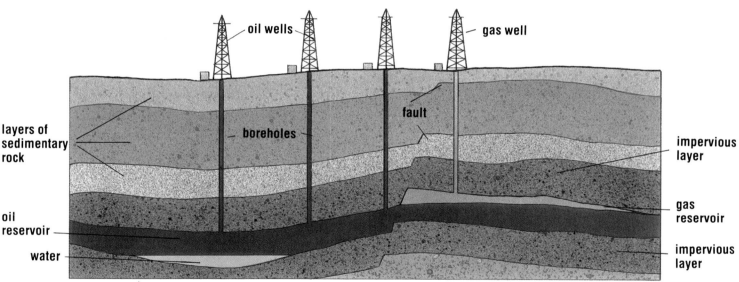

oil wells   gas well

layers of sedimentary rock   boreholes   fault   impervious layer

oil reservoir   gas reservoir

water   impervious layer

we tap them by drilling and make use of them as fuel.

These three fossil materials – oil, gas, and coal – are valuable as fuels because they are made up of carbon and carbon compounds. They combine readily with the oxygen of the air when they are ignited, and plentiful heat is given out. All three are also useful as rich sources of organic chemicals, which can be transformed into such products as plastics, synthetic fibers, pesticides, and drugs.

Because these fuels are fossils, their supply is limited. Nature is not replenishing them at the rate at which we are consuming them. Worldwide, we use something like 1,585,000,000 gallons of oil a day. At this rate of consumption, the oil wells will begin to run dry early next century, and natural gas will not last for much longer. Coal, however, is plentiful enough to last for many years.

So it seems that coal, the fuel that thrust humankind into our present technological age in the last century, will be instrumental in our continuing industrial expansion next century. Unless, of course, alternative energy sources can be developed.

Above: Oil and gas have become trapped in various rock formations underground. They travel through porous rocks until they encounter impervious layers. There, they become trapped. Holes drilled through the rock layers might strike oil, gas — or nothing. Often the oil or gas travels to the surface under pressure; sometimes it has to be pumped out.

ENERGY FROM THE ATOM

Above: The nuclear power station at Oldbury, England. Like all such power stations, it uses the fission of uranium atoms to produce heat.

Below: The layout of a typical nuclear power plant. The most important unit is the reactor, in which the uranium undergoes fission. A cooling fluid, or coolant, circulates through the reactor and extracts heat. This heat is used to raise steam to drive the steam-turbine generators that produce electricity.

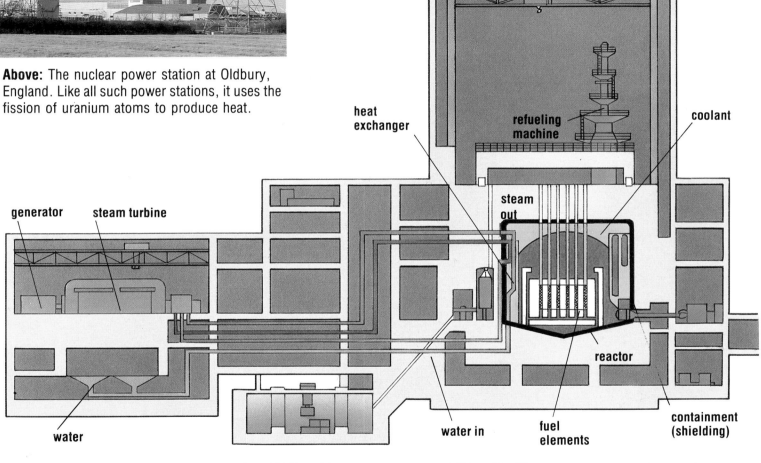

Right: Workers handling radioactive materials by remote control. The materials are kept in a sealed room with thick walls and windows that prevent dangerous radiation getting through and harming workers. Every worker in a nuclear plant wears a device called a dosimeter, which registers the amount of radiation they receive.

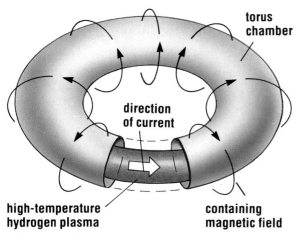

Above: Nuclear energy can also be released by the fusion, or joining together, of the nuclei of light atoms like hydrogen. This same process provides the energy that keeps the stars shining. Scientists carry out fusion experiments in electromagnetic devices like this torus chamber.

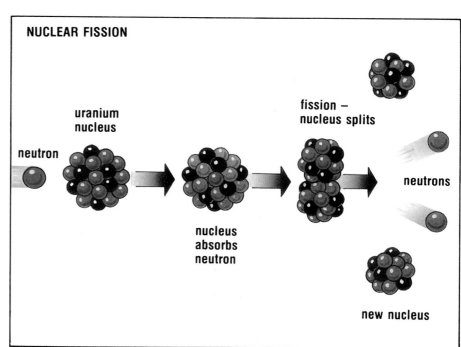

NUCLEAR FISSION

Above: The process of fission, which releases vast amounts of energy from the uranium atom. The process begins when a stray neutron (an electrically neutral particle found in most atoms) hits a uranium nucleus. The nucleus captures it, but then becomes unstable and splits in two. This fission process produces energy, which is given off as heat, light and radiation. Also given out are several neutrons, which can then go on to bring about fission in more uranium atoms. If this happens rapidly, a chain reaction can build up which results in the release of fantastic amounts of energy.

When oil and gas start to run out next century, the world will begin to suffer a severe energy shortage. Some people believe that the only answer will be to use more nuclear power. This utilizes energy locked within the nucleus of atoms, specifically the atoms of uranium, and a few other heavy metals.

Energy is released from the uranium atom when it is made to undergo fission or split. A fantastic amount of energy can be unleashed if millions of atoms are split in quick succession. If such a chain reaction takes place in, say, a pound of uranium, it will release as much energy as setting off tens of thousands of tons of a high explosive such as TNT.

It was in the form of an explosion that nuclear energy was first introduced to the world at large, in August 1945. An American bomber dropped an atomic bomb on the Japanese city of Hiroshima. It devastated the city and killed or wounded 150,000 people.

It was to be some years before nuclear energy was tamed, and the first nuclear power station to produce electricity was opened in December 1951 in the U.S.A. Britain's earliest was Calder Hall in Cumbria in 1956.

The heart of a nuclear power plant is the reactor, which contains the uranium "fuel." In the reactor, the uranium is allowed to undergo a controlled chain reaction. So-called control rods are moved in and out of the reactor core to regulate the output of energy. The energy is released mainly as heat and is carried from the core by a coolant, usually gas or water under pressure. The coolant gives up its heat in a heat exchanger, in which water is boiled into steam. The steam then drives turbine-generators to produce electricity.

The major drawback of nuclear power is that the fuel, the process, and the waste products from the reactor are highly radioactive. They give out penetrating radiation, that even in quite small doses, can kill. So elaborate precautions must always be taken to prevent the leakage of radiation.

ENERGY ALTERNATIVES

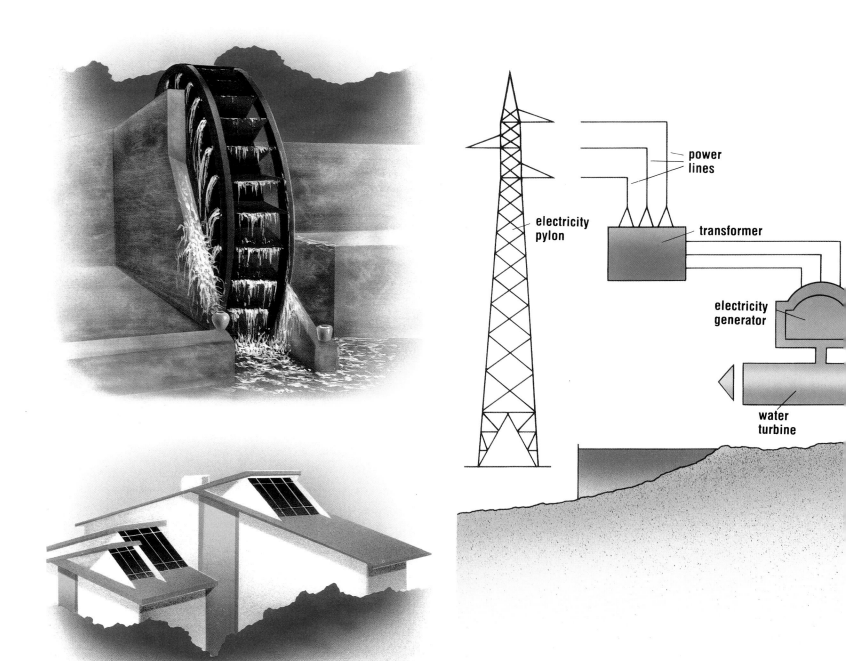

Top: The water wheel remained a major source of power for driving machines until recently.
Above: This house is fitted with large solar panels, that capture the Sun's heat.
Main picture: In hydroelectric schemes, water stored behind a dam spins the turbine and the electricity generator coupled to it. The electricity is then transmitted cross-country by power lines.

With fossil fuels running out and nuclear power being dangerous, we have to find other means of supplying the world with the energy it needs. There are several ways, and some are already being exploited, although it is doubtful whether they will be able to supply enough for all our future needs.

The most widely used form of "alternative energy" is water power. The Romans first harnessed the power of running water 2,000 years ago by means of waterwheels. Today, engineers use the modern

equilavent of the waterwheel, the water turbine. They use it to spin generators to make electricity. Electricity generated in this way is called hydroelectric ("water-electric") power. About one-third of the world's electricity supply is hydroelectric.

In most hydroelectric schemes, water is stored behind a dam. Then, it is released through turbines installed at a lower level. The depth, or "head," of water, provides the force to drive the water through the turbines. In

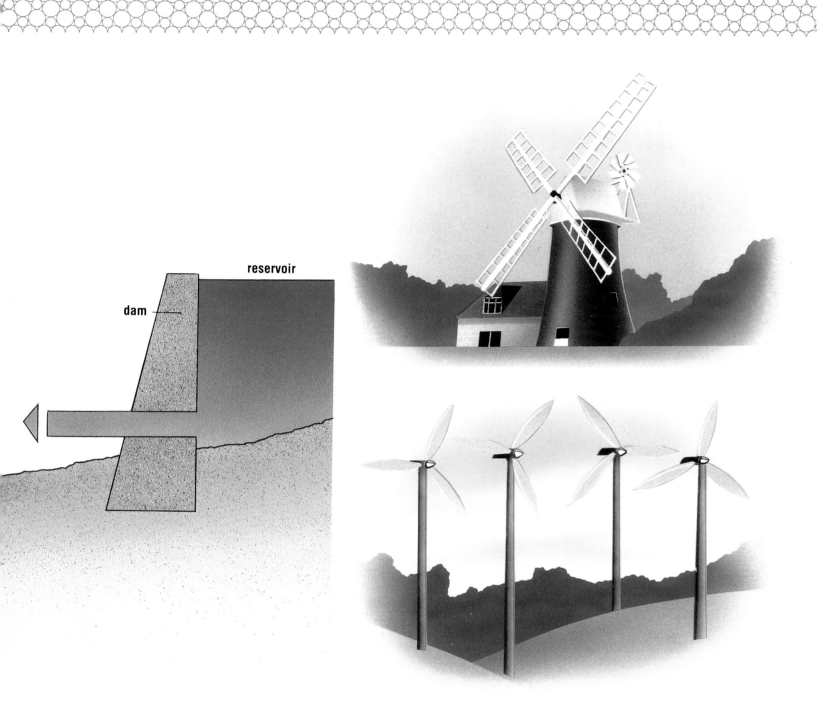

some water-power schemes, the rise and fall of the tides provide the "head" for the turbines. The best-known tidal power scheme is located across the River Rance in Brittany, France.

Another old idea – wind power – has also been revived. Modern wind turbines, the equilavent of the old windmills, harness the power in the wind to spin electric generators. In some areas, notably in California, vast wind "farms" have been built to feed electricity into the local grid.

The winds are themselves driven by energy from the Sun. But we are also now beginning to harness the Sun's heat directly, in a number of solar energy projects. The most ambitious are the so-called solar power towers. They use banks of mirrors to reflect sunlight onto a tower, where the concentrated heat is used to boil water into steam to drive electric generators. At the residential level, solar panels are able to supply hot water and heating even in relatively cool climates.

Top: Windmills like this once dotted the countryside and were used for grinding, or milling, grain into flour. The wind turned the wooden sails of the mill, and the movement was transmitted to the grindstones by shafts and gear wheels.
Above: Propeller-type wind turbines on a modern wind "farm." The spinning propellers drive electricity generators.

ENGINES AND TURBINES

Right: A typical gasoline engine, which has four cylinders and works on the four-stroke cycle (bottom picture). The valves let fuel into, and exhaust gases out of, the cylinders opened by a rotating camshaft. The crankshaft converts the up-and-down movement of the pistons into rotary motion, which turns the heavy flywheel.

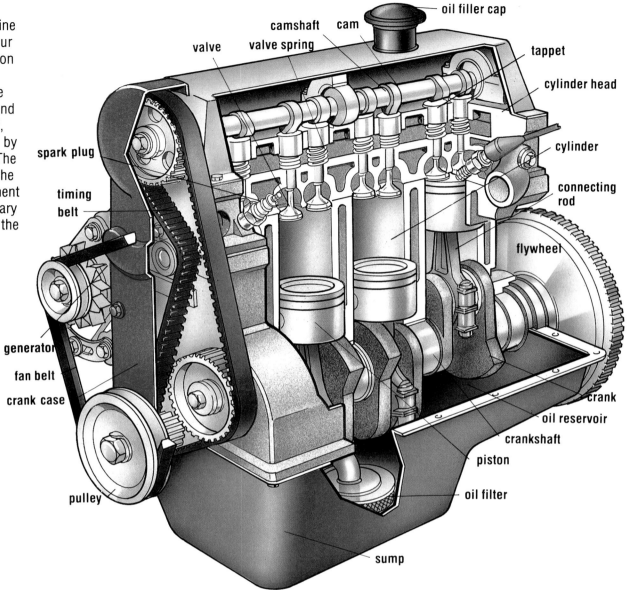

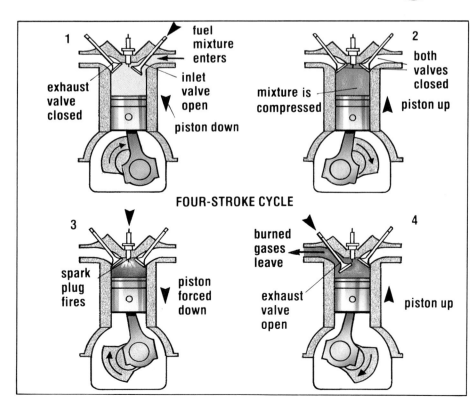

FOUR-STROKE CYCLE

1. exhaust valve closed — fuel mixture enters — inlet valve open — piston down

2. mixture is compressed — both valves closed — piston up

3. spark plug fires — piston forced down

4. burned gases leave — exhaust valve open — piston up

Engines and turbines are machines that put energy to work. Our most familiar engines are heat engines, which burn fuels to produce heat. This input of heat causes a gas to expand and, in gas and diesel engines, move a piston; or in turbines, spin a blade.

In gas and diesel engines, the up-and-down movement of pistons in cylinders is converted to a rotary motion by the use of a crankshaft. This rotary motion is then able to turn the driving wheels of vehicles, for example.

In a gas engine, the first downstroke of the piston draws a gas/air mixture into the top of the

Right: A hovercraft ferry, which transports passengers and vehicles across the English Channel. The hovercraft glides over the surface of the water on top of a cushion of high-pressure air. The air is forced underneath the vehicle by powerful fans, driven by gas-turbine engines. These engines also spin the propellers that move the hovercraft.

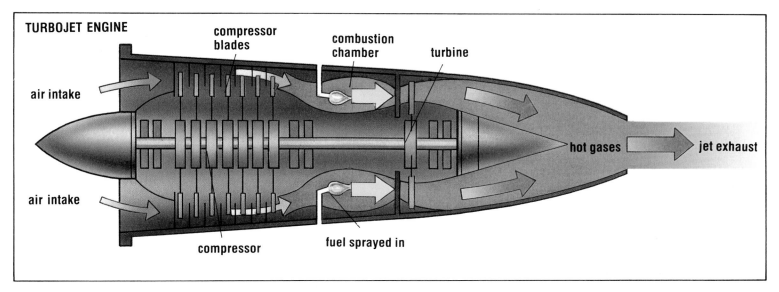

TURBOJET ENGINE

compressor blades

combustion chamber

turbine

air intake

air intake

hot gases

jet exhaust

compressor

fuel sprayed in

cyclinder. On the next stroke up, the piston compresses the mixture, which is exploded by an electric spark. The hot gases produced expand and force the piston down, producing power. Next, the piston comes up again, forcing the burned gases out of the cylinder. This cycle is then repeated.

In the diesel engine, the cycle is different. On the first downstroke, only air is taken in. This is then compressed so much that it gets very hot. Fuel is then injected into this hot air, and immediately explodes, driving the piston down.

Turbines produce rotary motion directly as gas, steam, or water, rush through their blades. In a gas turbine, hot gases are produced by burning fuel in high-pressure air from a compressor. The gases spin the turbine wheels as they leave the engine. One of the turbines turns the compressor.

A jet engine is a kind of gas turbine designed to allow as much gas as possible to escape backwards out of the engine. This jet of gas provides the thrust to propel aircraft. The jet engines fitted to airliners have huge fans in front of the compressors and are known as turbofans. Turboprops are jet engines that have a propeller spun by one of the turbines.

Above: A turbojet, the simplest type of jet aircraft engine. Air is sucked into the engine by a rapidly rotating compressor fitted with many stages, or sets of blades. The compressed air is channeled into the combustion chamber. Kerosene fuel is sprayed into the compressed air and ignited. The hot gases produced spin the turbine before escaping from the engine as a high-speed jet.

STATIC ELECTRICITY

Above: Lightning occurs after static electricity has built up in thunder-clouds. The electricity reaches such high voltages (millions of volts) that it charges atoms in the air. These charged atoms provide a path to conduct the lightning both between the clouds and also down to the ground.

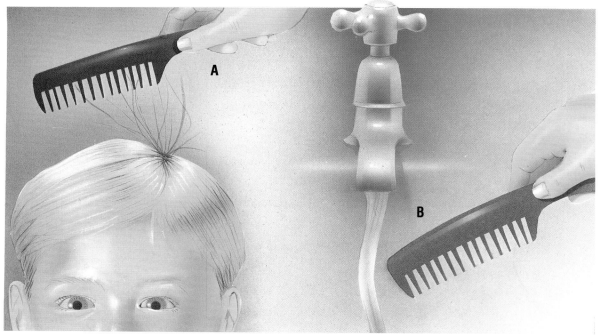

Sometimes when you walk across a carpeted room and then touch a metal door handle, you get an electric shock. When you take off a nylon shirt or blouse, you often hear it crackle. The shock and the crackling occur because you and your clothing have been electrified. They have acquired an electric charge, which builds up.

The shock and the crackling happen when the electricity suddenly discharges, or leaves. When it leaves through your fingertips, you feel the prickle of an electric shock. When it leaves your clothes, the electricity discharges as tiny sparks, which cause the crackling.

This kind of electricity is known as static electricity. It is so called because it tends to stay where it is produced and does not flow through metal wires like the electricity produced by a battery. Indeed, you cannot produce static electricity in metals. You can produce it only in insulators – materials that do not conduct (pass on) electricity. Examples of insulators include plastics, rubber, and glass.

These materials can be given an electric charge by rubbing. The rubbing, or friction, of your feet on a plastic (synthetic-fiber) carpet produces enough charge to give you a shock. The rubbing of, say, a wool sweater against a nylon blouse builds up enough charge to create sparks. Rub a balloon against your sweater, and the balloon will become charged. You will be able to "stick" it on the ceiling.

Comb your hair with a plastic comb, and both will become charged. Hold the comb above your head, and your hair will stand on end. Your comb and hair have acquired different electric charges (+ and −), and these charges attract one another.

The most spectacular display of static electricity, however, occurs in the atmosphere. The rubbing together of droplets and particles in a rain cloud produce very high electrical charges. When this electricity discharges, we see the great sparks we call lightning.

Below: The pictures at the bottom of the page show some tricks you can play with static electricity. (**A**) By combing your hair with a plastic (not a metal) comb, you can make your hair stand on end. (**B**) Comb your hair again, and hold the comb next to a stream of water from a tap. What happens? (**C**) Blow up a balloon and rub it against your sweater to charge it up. (**D**) Do this with two balloons and then hang them close together. What happens? (**E**) Place a charged-up balloon against the wall or ceiling. Does it fall or defy gravity?

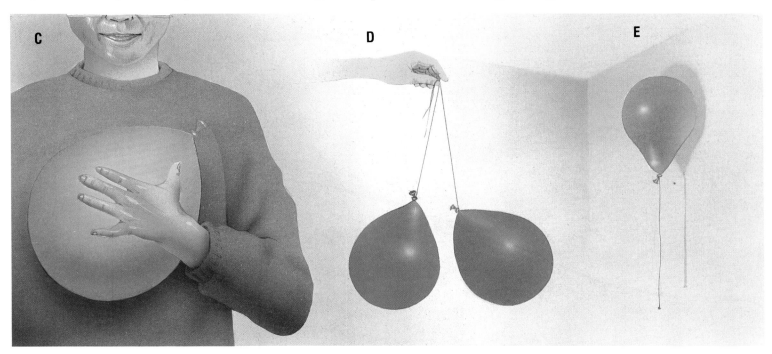

C   D   E

CURRENT ELECTRICITY

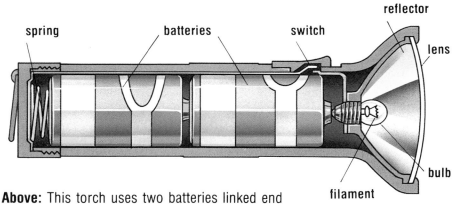

spring   **batteries**   **switch**   **reflector**   **lens**   **bulb**   **filament**

Above: This torch uses two batteries linked end to end. One terminal of the bulb is in contact with the positive terminal of the top battery. Pressing the switch completes the circuit between the negative terminal of the bottom battery and the other terminal of the bulb, which then lights up.

positive electrode (carbon)   **negative electrode (zinc)**

chemical paste

Left: A look inside a dry cell (battery). Chemical reactions take place between the two electrodes (carbon and zinc) and the chemical paste between them. A flow of electrons takes place between the two electrodes when they are connected by an external circuit.

Below: In a simple electric motor (left), current from a battery passes through the field coils, turning them into electromagnets. As current flows through the armature coils, the armature is forced to turn. The generator (right) works like a motor in reverse, producing current in the armature coils when it is rotated.

Flowing, or current, electricity, is much more useful than static electricity. It powers our flashlights, radios, cassette players, house lights, stoves, and appliances of all kinds. Batteries are a portable source of electric current. They produce electricity by means of chemical reactions. These reactions produce a flow of electrons (see page 20) between the battery terminals, or electrodes. This flow constitutes an electric current.

In an ordinary dry-cell flashlight, reactions take place in a chemical paste when the two terminals are connected by an external path, or circuit. The positive terminal is a carbon rod with a brass cap; the negative terminal is the zinc casing of the battery. We say in science that the electric current flows from positive to negative. However, in practice, the electrons flow in the circuit from the zinc to the carbon rod (negative to positive).

The electricity we use at home is produced in quite a different way.

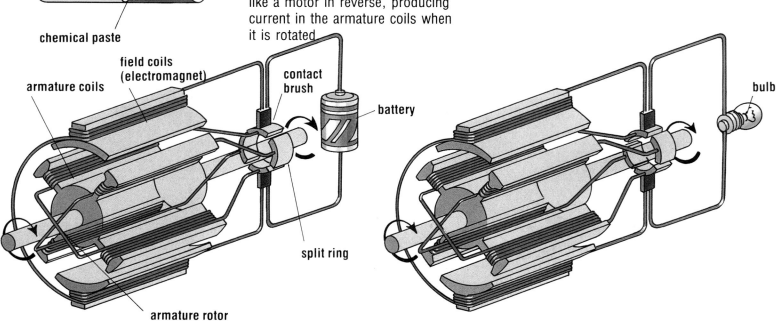

armature coils   **field coils (electromagnet)**   **contact brush**   **battery**   **bulb**   **split ring**   **armature rotor**

It is produced by powerful machines called generators. They work on a principle of electro-magnetism (see page 38) known as the dynamo effect. When you move a wire in a magnetic field, you set up an electric current in it. A practical generator uses huge coils of wire wound on a shaft (rotor) that rotates inside the poles of a powerful electromagnet. The rotor is spun by a coupled turbine, driven by steam or flowing water.

The "pressure" of electricity is measured in units called volts. A flashlight battery has a voltage of only 1.5 volts, while generators produce electricity at 25,000 volts. By the time it gets into our homes the electricity has been reduced to about 120 volts.

As it flows through coils in the stove, the electricity is turned into heat. As it flows through the filament of electric light bulbs, it makes them glow white-hot and give out light. Electricity is dangerous and can deliver a shock that can kill.

Substations reduce the voltage to the level that different consumers require. The voltage is increased and reduced by transformers.

boiler house

steam turbines

generator

power station

power lines

pylon

substation

"step-up" transformer

"step-down" transformer

electrified railroad

Mains electricity is often produced at coal-fired power stations (top). They burn coal to produce heat to raise steam in their boilers. The steam is used to spin steam turbines, which are coupled to the generators that produce electricity.

"step-down" transformer

heavy industry

The generated electricity is first increased greatly in voltage, and then sent along power lines which are carried by pylons to substations.

light industry

"step-down" transformer

farmhouse

village

city

"step-down" transformer

MESSAGES BY WIRE

If you wanted, you could send coded messages to someone inside their house, by ringing their doorbell. If you both knew Morse Code, you could send messages as dots (short rings) and dashes (long rings). An apparatus similar to a doorbell pioneered long-distance communications more than 150 years ago. One of the great scientific pioneers was an American named Samuel Morse (1791-1872), who devised the dot-and-dash Morse Code.

This method of communication depended on electric current traveling through wires. It was called the electric telegraph.

In every telegraph, the signal was produced in the receiver when electric current passed through an electromagnet. As it was energized, the electromagnet attracted a piece of metal, which activated an inked pen, caused a click, or triggered off a buzzer.

The monopoly of the telegraph for long-distance communications did not last long. In 1876, another American, Alexander Graham Bell, invented the telephone. Now it became possible for people to send spoken messages along the wires. Today the telegraph is virtually dead, although some businesses still use a modern version, called a telex, for some communications.

The telephone, like the telegraph before it, uses electric current through wires to carry messages. So the telephone receiver – the part you pick up – incorporates devices to convert sound waves into electric signals and vice versa.

The device in the telephone mouthpiece is a kind of micro-

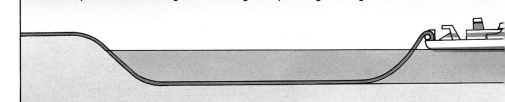

Below: Telecommunication cables are laid underwater as well as on land and may link continents. A successful telegraph cable was laid across the Atlantic in 1866, but the first telephone cable was not laid until 90 years later. Nowadays, however, most intercontinental telecommunications are routed via communication satellites. Submarine cables are well insulated and incorporate electronic devices called repeaters to strengthen the signals passing through.

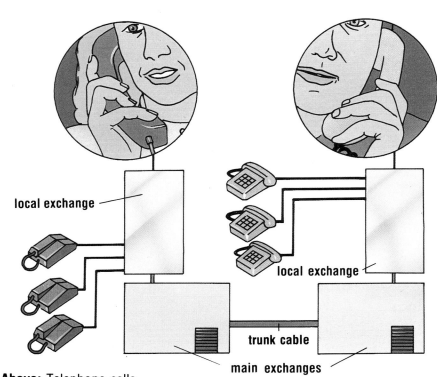

local exchange

local exchange

trunk cable

main exchanges

Above: Telephone calls are routed through a network of exchanges. Each phone is connected to a local exchange, which in turn feeds calls into a main exchange.

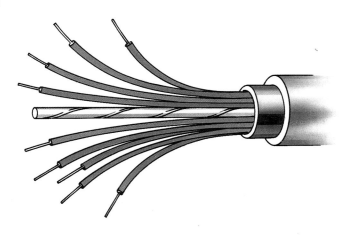

Right: Increasingly, telephone engineers are laying optical-fiber cables (made of fine glass fibers) in place of copper ones. Signals are sent along them on laser beams.

phone. It has a thin disk, or diaphragm, that vibrates when you speak. The vibrations are made to alter an electric current, so setting up varying signals. The earpiece also contains a diaphragm. This is made to vibrate and produce sound waves when the varying electric signals are passed through an electromagnet underneath.

Words and pictures as well as voices are now sent along telephone wires by means of fax. A fax, or facsimile machine, converts printed words or other images into a pattern of electric signals, which are then transmitted. They are converted back into images by the receiving machine.

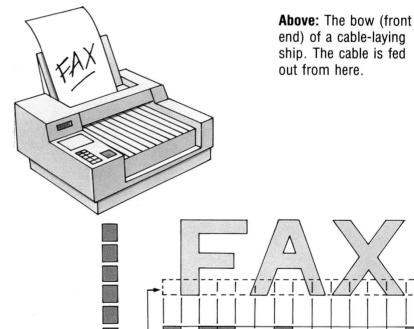

scanned line

Right: The past few years have seen an enormous increase in the use of facsimile, or fax machines. They transmit printed matter as electrical signals along ordinary telephone lines. The transmission of an A4 page takes only about a minute. In the machine shown, the document to be transmitted feeds through the machine from back to front. In some machines, the paper feeds through in the opposite direction.

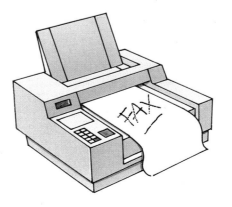

Above: The bow (front end) of a cable-laying ship. The cable is fed out from here.

Above: In a fax machine, the printed matter is read line by line by a scanning device. It converts the pattern of light and dark in each scanning line into a series of electric charges, which set up the electrical signals that are transmitted down telephone lines.

MESSAGES ON THE AIR

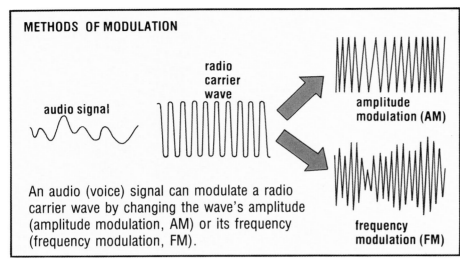

METHODS OF MODULATION

audio signal

radio carrier wave

amplitude modulation (AM)

frequency modulation (FM)

An audio (voice) signal can modulate a radio carrier wave by changing the wave's amplitude (amplitude modulation, AM) or its frequency (frequency modulation, FM).

The majority of overseas telephone calls these days travel via communications satellites in space, 22,400 miles above the Earth. They do not reach the satellites via long wires, but by means of microwaves – short radio waves. Experiments in radio communications began at about the turn of this century, with the Italian Guglielmo Marconi doing pioneering work in England. The first regular radio broadcasts began in the 1920s.

Radio works by transmitting signals on radio waves, which are electromagnetic waves related to light, but with a much longer wavelength. The sounds of voice and music are first converted into electrical signals by a microphone. Then these signals are made to change, or modulate, radio waves.

The modulated radio wave travels to the aerial of a radio receiver. Electronic circuits in the receiver demodulate, or separate out, the voice signals and send them to a loudspeaker, which converts them back into sound.

Right: Various kinds of waves are used for radio transmissions. Short and medium waves travel a long way because they are reflected by layers in the atmosphere. Microwaves are used to transmit signals to satellites and relay towers.

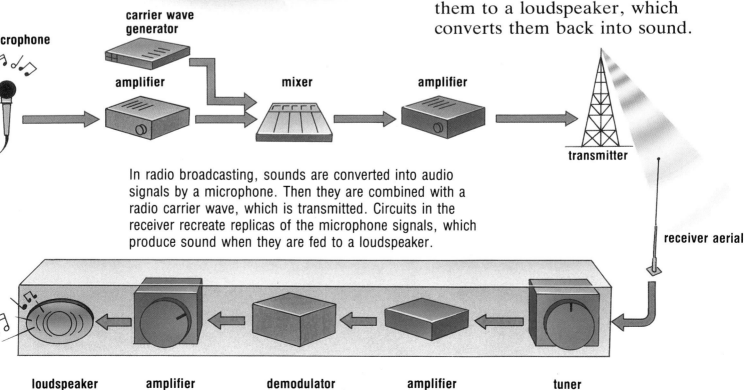

communication satellite

reflecting layers in ionosphere

microwaves

short radio waves

relay tower

carrier wave generator

microphone

amplifier

mixer

amplifier

transmitter

receiver aerial

In radio broadcasting, sounds are converted into audio signals by a microphone. Then they are combined with a radio carrier wave, which is transmitted. Circuits in the receiver recreate replicas of the microphone signals, which produce sound when they are fed to a loudspeaker.

loudspeaker   amplifier   demodulator   amplifier   tuner

Pictures as well as sound can be transmitted by radio waves, which allows the phenomenon of television. In a TV camera, electronic tubes convert the pattern of light entering the camera lens into patterns of electric charges. Traveling electron beams scan the patterns in a series of lines.

The vision signals are carried by radio waves to the aerial of a TV receiver. This houses a tube with a fluorescent screen. The tube generates a beam of electrons, which scans the screen in lines and makes the screen glow where it strikes. The vision signals are separated from the radio carrier wave and are made to vary the brightness of the scanning beam. This recreates the pattern of light in the scene viewed.

Color is created by a pattern of dots (chemicals called phosphurs) on the fluorescent screen. They react with the electrons shot toward them from electron guns to glow either red, blue, or green. The colors merge on the screen to give the color of the original image.

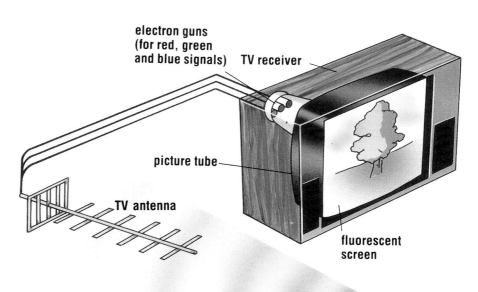

Above: In the receiver, the original pattern of colored light is recreated line by line by electron guns. They fire electrons at phosphor dots on the TV screen, which glow red, green, or blue, as appropriate.

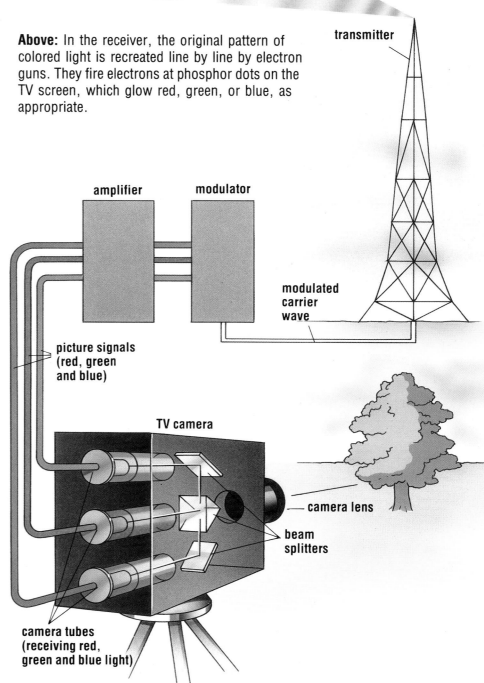

Right: In a color TV camera, three electronic tubes (vidicons) record the patterns of red, green, and blue light in the scene viewed by the camera lens.

Right: The camera tubes generate red, green, and blue picture signals for each line scanned by an electron beam. They are then transmitted.

ELECTRONS AND CHIPS

Left: A technician putting the finishing touches to a flexible circuit board, widely used in the electronic industry. Practically all electronic devices, from digital watches and radios to television sets and computers, are built up from self-contained modules whose components are mounted on circuit boards.

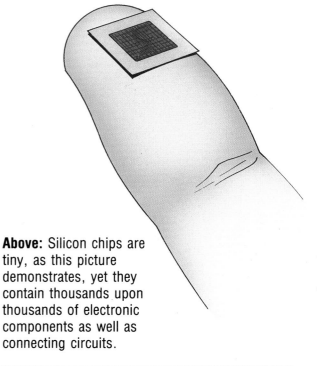

Above: Silicon chips are tiny, as this picture demonstrates, yet they contain thousands upon thousands of electronic components as well as connecting circuits.

Below: This simple pocket calculator has number and function keys, which are pressed to bring about calculations. The sequence of steps needed to perform a simple calculation is shown. Pressing a number key causes the number to appear in the display, which is a liquid crystal display (LCD). Figures are formed when separate segments of liquid crystal turn black.

liquid
crystal
segments

liquid
crystal
display

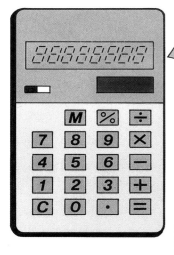

keys used
to input
first number

key used
to input
operation

keys used
to input
second number

key used
to obtain
result

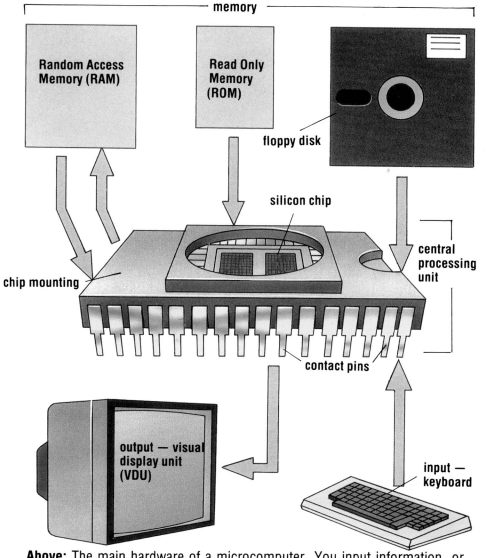

memory

Random Access Memory (RAM)

Read Only Memory (ROM)

floppy disk

silicon chip

central processing unit

chip mounting

contact pins

output — visual display unit (VDU)

input — keyboard

Above: The main hardware of a microcomputer. You input information, or data, into it, and communicate with it through the keyboard. The computer operates according to instructions stored in a permanent read-only memory (ROM). It stores programs fed into it from a floppy disk, as well as data currently being worked on in a temporary random-access memory (RAM). The central processing unit (CPU), which contains a microchip, processes the data and outputs the results to the computer screen, or visual display unit (VDU).

Right: In the binary number system used by computers, place values go up in powers of 2. The table shows the binary numbers equivalent to the decimal numbers 1-10. The tape is punched in binary code.

Decimal Number ↓ →	Binary numbers			
	2^3 (=8)	2^2 (=4)	2^1 (=2)	2^0 (=1)
1				1
2			1	0
3			1	1
4		1	0	0
5		1	0	1
6		1	1	0
7		1	1	1
8	1	0	0	0
9	1	0	0	1
10	1	0	1	0

Punched tape

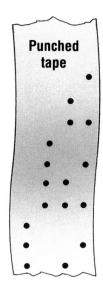

Tiny slivers of crystal, the size of a thumbnail, have ushered in a revolution in business, industry, leisure, and learning. These slivers are silicon chips. They are the driving force behind the electronics revolution that is transforming our lives. They are the "brains" behind the digital watch, pocket calculator, microcomputer, and the powerful large mainframe computer.

Computers large and small work in much the same way. Data (information) is fed into them from a keyboard or a magnetic disk, or tape. A program (set of instructions) tells the computer to process the data in some way, for example, to sort it or carry out calculations. The program is stored in the computer's memory. Other data may be stored in another part of the memory. Finally, the program tells the computer to display the results on a VDU (visual display unit, or screen); print them out on paper; or store them on disk or tape.

The computer cannot process data or instructions in ordinary figures or words. It can only operate when they are translated into binary code. This uses just two values, 0 and 1, the so-called binary digits (bits). Fortunately, the computer itself translates words and figures into binary code when the keyboard is pressed.

Computers can do much more than compute. They can be used to play chess and other exciting games; create graphics, or images, in color, and even in 3D (three dimensions); and act as word processors to create texts that can be edited, filed, converted, replaced, and merged.

WHAT IS LIGHT?

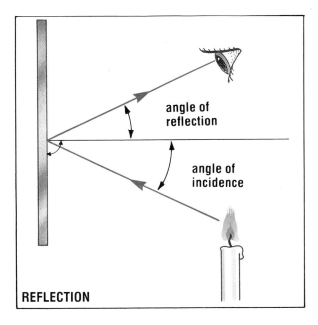

angle of reflection

angle of incidence

REFLECTION

Left: We see objects because they reflect light into our eyes. A mirror has a shiny silvery surface that reflects light perfectly. When a light ray is reflected by a plane (flat) mirror, it leaves the mirror at the same angle as it approached.

Right: The periscope is a useful optical instrument made using two plane mirrors.

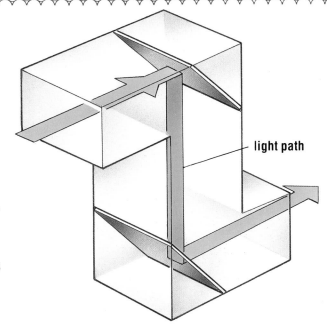

light path

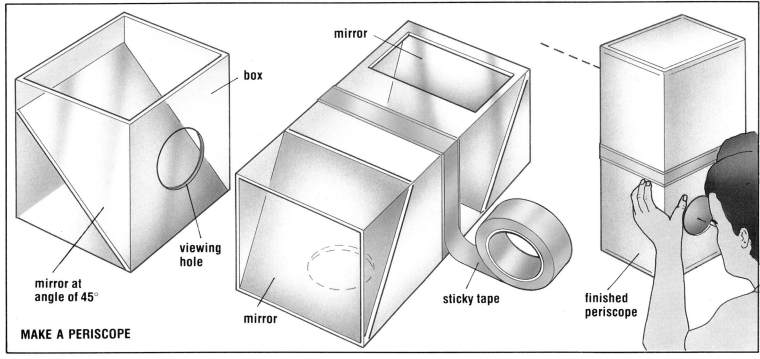

mirror

box

viewing hole

mirror at angle of 45°

mirror

sticky tape

finished periscope

MAKE A PERISCOPE

Above: You can make a periscope yourself, and use it to look over walls or around corners. Fit two mirrors in boxes, as shown, so that they are inclined exactly at an angle of 45 degrees. Cut holes for viewing. Tape the two boxes together (make sure they are the right way around!), and your periscope is ready to use.

Of the five senses – sight, hearing, taste, smell, and touch – sight is arguably the most important, at least to human beings. We see things because our eyes are sensitive to light. During the day, most of the light in our world comes from the Sun. We see objects because they reflect sunligth into our eyes.

Light travels in straight lines, made up of many wavelengths (see page 66). If it traveled in a curved path, you would be able to see around corners, and you would not get sharp shadows. Light travels readily through the air, and through glass and water. We say that these media are transparent to light. Materials through which light cannot travel, such as wood and rock, are called opaque.

These materials reflect light, but not well. Polished metal, however, reflects light nearly perfectly. Most mirrors are made by covering one side of a glass plate with a silvery metal coating.

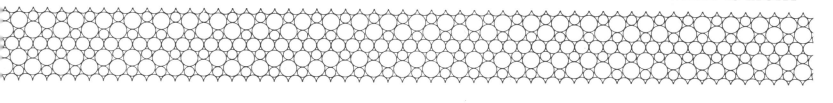

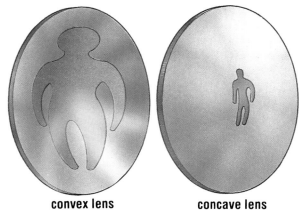

convex lens   concave lens

Left: When you look in an ordinary plane (flat) mirror, you see a mirror image of yourself, with left and right reversed. But when the mirror surface is curved, your image is distorted. Halls of mirrors at carnivals use combinations to produce funny effects.

Right: A convex lens bends in a beam of parallel light rays to a focal point. A concave lens spreads out a beam of parallel light rays.
Below: Look at a coin in a glass of water from the top. It appears to be much closer to the surface than it really is. This is due to refraction.

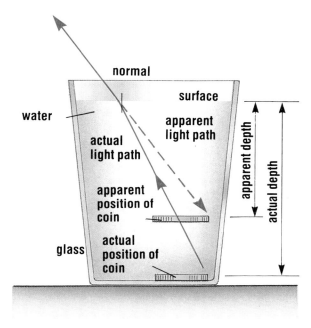

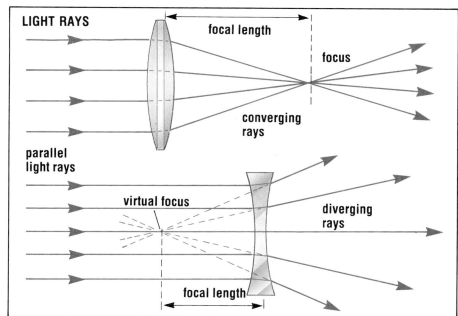

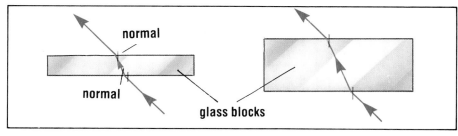

We think of light as traveling instantly from place to place. And in our everyday life, this is perfectly valid, for light travels at a speed of nearly 186,500 miles per second. So in only one second it could travel around the Earth nearly seven times!

Light travels slightly slower in water and glass. This causes a ray of light to bend slightly when it enters these media. We call this bending of light refraction. Refraction makes a stick floating half out of the water look broken, and a swimming pool look shallower than it really is.

We use the principle of refraction to make glass lenses. These have curved surfaces and bend light in different ways. Convex lenses are lenses that bulge in the middle. Concave lenses go in at the middle. Convex ones are the most useful, because they can be used to focus light and form images; they can also be used to produce a variety of magnified images.

Above: A ray of light bends as it travels from one medium to another. When it travels from a less dense to a more dense medium, say from air into glass, it is bent toward the normal. When it travels from a more dense to a less dense medium, say from glass into air, it is bent away from normal.

SEEING NEAR AND FAR

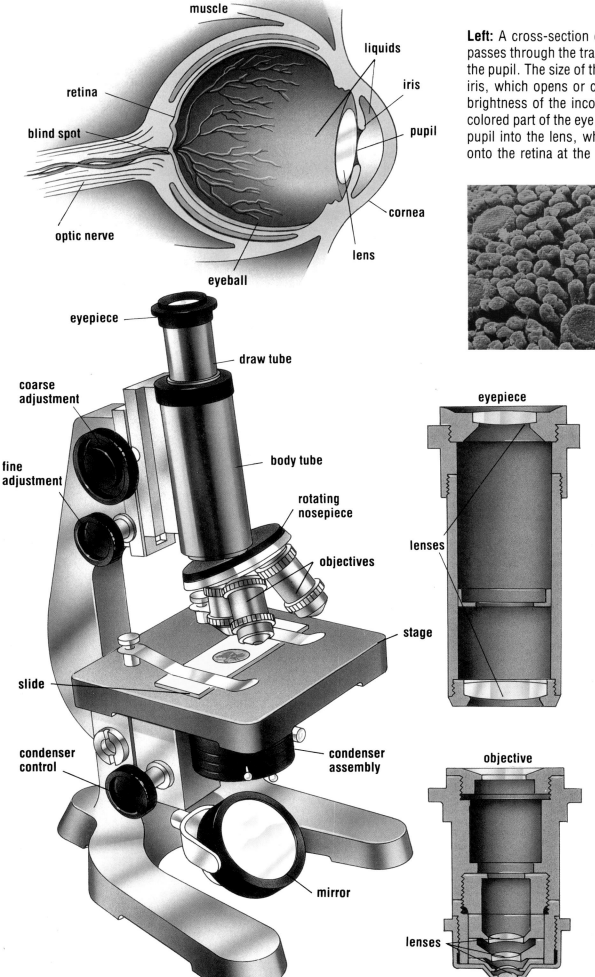

muscle

liquids

iris

pupil

retina

blind spot

cornea

optic nerve

lens

eyeball

Left: A cross-section of the human eye. Light passes through the transparent cornea and enters the pupil. The size of the pupil is governed by the iris, which opens or closes according to the brightness of the incoming light. The iris is the colored part of the eye. Light passes through the pupil into the lens, which focuses the light rays onto the retina at the rear of the eyeball.

eyepiece

draw tube

coarse adjustment

fine adjustment

body tube

rotating nosepiece

objectives

slide

stage

condenser control

condenser assembly

mirror

eyepiece

lenses

objective

lenses

Above: This false-color picture of the surface of a human tongue was taken by an electron microscope. This instrument produces magnified images of objects using beams of electrons rather than light rays. Its "lenses" are electromagnetic coils rather than pieces of curved glass.

Left: An ordinary light microscope, which is a compound microscope with two sets of lenses. The specimen to be examined is placed on a slide and set up on the stage. Light is reflected through the specimen via a mirror and condenser lens beneath the stage. One of the objectives on the turret, or nosepiece, is moved into position, and then the tube containing the eyepiece is moved in or out to bring the specimen into focus. The different objectives in the nosepiece have different powers of magnification.

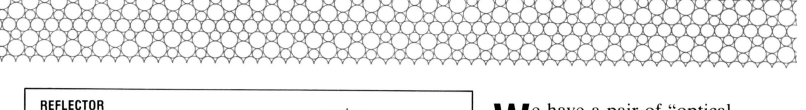

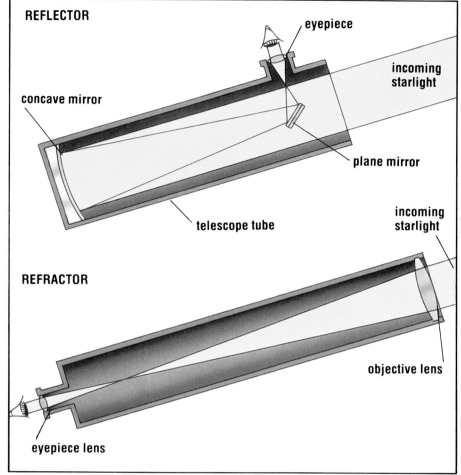

Above: Two different kinds of telescopes used by astronomers to view the skies. The refractor uses two sets of lenses to gather and focus the light from the stars. The reflecting telescope uses a concave mirror to do so. The type illustrated is called a Newtonian reflector. The curved mirror reflects light onto a plane mirror near the top of the telescope tube, and this mirror in turn reflects it into an eyepiece set in the side.

Below: This photograph of an open star cluster was taken through a reflecting telescope with a mirror over 13 feet in diameter.

We have a pair of "optical instruments" in our head – our eyes. Light rays enter the eyeball through a transparent layer (cornea), and an opening, or aperture (pupil) lets them into a lens. The lens focuses the light and forms a sharp image on a screen (retina) at the rear of the eyeball. Light-sensitive cells record the image and send signals to the brain, which interprets them as a picture.

If the lenses in your eyes are defective, they may not be able to form a sharp image on the retina. Then you must place additional lenses in front of your eyes, as glasses or contact lenses, to bend the incoming light so that your own lenses can form a sharp image.

The most common optical instrument apart from the eyes is the camera, with which we take photographs. It is constructed much like the eye. It is a light-tight box, which light enters through a lens. The lens forms an image on a piece of film. A shutter normally covers the film, but it is opened for a fraction of a second when a picture is taken.

As mentioned earlier (page 63), convex lenses can produce magnified images. This property is used in microscopes and telescopes. These instruments use two sets of lenses to see tiny things (microscope) or distant objects (telescope). One lens (objective) forms a magnified image, and then the second lens (eyepiece) magnifies it even more. All the biggest telescopes, however, use curved mirrors instead of lenses to gather and focus the faint light coming from the stars.

COLORS AND WAVES

We see flowers, neon lights, clothes, and other painted or dyed materials from the natural and human world in all the colors of the rainbow – and a few more.

The main colors of the rainbow are violet, indigo, blue, green, yellow, orange, and red. But where do those colors in the sky come from? The answer is, they come from sunlight. To us, sunlight appears white, but this masks its true nature. When sunlight travels through raindrops, it does not emerge as white light, but as a spread of colors – the colors of the rainbow. We call this spread of color a spectrum.

This spectrum is produced because white light is actually a mixture of light of different wavelengths. Each wavelength has a different color and bends at a slightly different angle as it passes through the raindrop. The result is that the color spreads. We can produce a spectrum from white light by passing it through a prism.

The wavelengths of light are part of a much bigger spectrum, called the Electromagnetic spectrum. This includes the waves we know as gamma-rays, X-rays, and ultraviolet rays, which have a shorter wavelength than light. Waves with a longer wavelength than light include infrared rays, microwaves, and radio waves.

All these different waves belong to the same family. They are electric and magnetic vibrations, which all travel at the speed of light. Their wavelength varies from a few million-millionths of a foot for gamma rays, to more than half a mile making up the longest radio waves.

Above: This fine double rainbow was photographed in the Pennine Mountains in England. The main, or primary, rainbow is most vivid, with red on the outside and violet on the inside of the curve. Notice that the colors are reversed in the weaker secondary rainbow.

Below: When ordinary white light is passed through a prism, it emerges as a band of rainbow color we call the spectrum. It comes about because the different wavelengths (colors) of light are bent (refracted) by different amounts at the air/glass and the glass/air surfaces. The red rays are bent the least and the violet rays the most; and it is this that causes the spread of color.

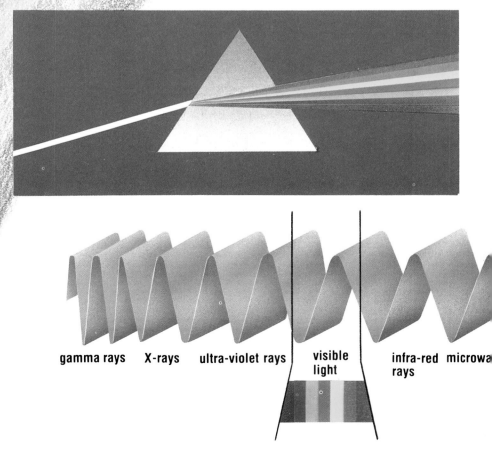

gamma rays   X-rays   ultra-violet rays   visible light   infra-red   microwa rays

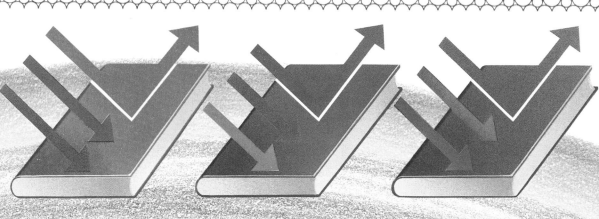

Right: Every object both absorbs and reflects the light falling on it. And it is selective in the wavelengths, or colors, it absorbs or reflects. An object that appears red looks that way because red is the only color that it reflects. It absorbs the other colors.

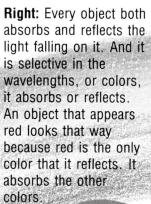

Right: A blackboard appears black because it absorbs all the colors in the light falling on it. None are reflected. But the chalk used to write on the blackboard reflects all the colors in the light that falls on it, which combine to produce white.

All colors can be produced by mixing together different amounts of red, green and blue light **(above)**. These are the primary colors of light. The primary colors in paint **(below)** are different.

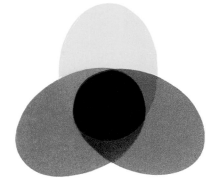

In ordinary white light, this picture of a red rose and green foliage **(above)** looks normal. But if you look at it in red light **(right)**, all you can see is the foliage in black: there is no green light to reflect.

Left: Visible light is only a tiny part of the electro-magnetic spectrum. The other rays, or waves, that make up this broad band are invisible to the eye. The illustrates how the waves differ in wavelength: shortest for gamma rays, longest for radio waves.

short radio waves

long radio waves

ELECTROMAGNETIC SPECTRUM

LASER LIGHT

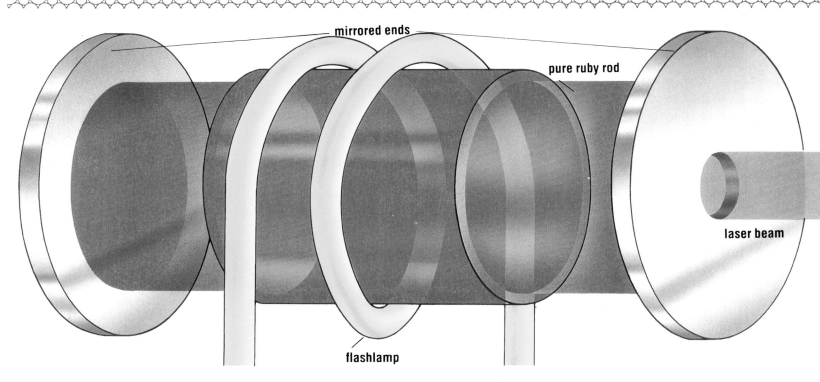

mirrored ends

pure ruby rod

laser beam

flashlamp

Above: A ruby laser. At each end, there are parallel mirrors. A flashlamp injects energy into the ruby rod when it is switched on. Atoms in the ruby gain extra energy, but then give it off as light of a certain wavelength, or color. This light triggers other atoms into giving off light of the same color. The light builds up in intensity as it is reflected back and forth by the mirrors. It comes out of one end as a very intense and parallel laser beam of pure light.

Above: Thin beams of laser light show up vividly in a laser display on Oxford Street in London. Laser beams are a concentrated source of light energy and hardly spread at all. They can even be reflected from the Moon and detected back on the Earth.

One of the purest kinds of light scientists can produce is laser light. It is light of a very precise wavelength, very intense, with all its waves exactly in step, and is produced in a beam that is almost perfectly parallel. Ordinary light, by contrast, is made up of many wavelengths, vibrating in all planes, with all waves out of step, and traveling in all directions.

Laser light is now widely used in industry, science, communications, medicine, and even in the home. In industry, it is used to cut and drill holes in metals. In science, it is used to investigate chemical reactions. In communications, it is used to transmit telephone messages through optic fibers. In medicine, it is used as a scalpel in surgery. And in the home, it is used to play compact disks.

One of the most exciting uses of lasers is in holography, a form of 3D (three-dimensional) photo-

Left: A boy is fascinated by a display of holography at the Museum of La Villette in Paris. The holographic image he is looking at has been created by laser light, and is truly three-dimensional. As the boy moves around the enclosure in which the image appears, he is able to see its front, sides, and back, just as he would if it were a real object.

Right: An industrial laser being used to cut and shape a metal workpiece. Laser light is produced in the form of an intense parallel beam. Like ordinary light, it can be focused by lenses. When it is focused, it produces an even more intense beam of concentrated energy. In industry this can be used for cutting metal, as here. In medicine lasers can be used instead of scalpels for cutting flesh in surgery.

graphy. Holographic images are most commonly used on credit cards. When you look at them from different angles, you see different aspects of the featured subject.

The word "laser" stands for "light amplification by the stimulated emission of radiation". The first lasers used a ruby rod to produce a laser beam, and the ruby laser is still widely used. It has accurately parallel ends, which are silvered to make them into mirrors.

Energy is fed into the rod as light from a powerful flashlamp. This gives atoms in the rod extra energy. Some radiate colored light as they lose this energy. This light also makes, or stimulates, other atoms to radiate the same colored light. The light is reflected back and forth by the mirrors, building in strength as more atoms are stimulated. Eventually, it emerges as an intense beam, powerful enough to cut through steel.

VARIETIES OF SOUND

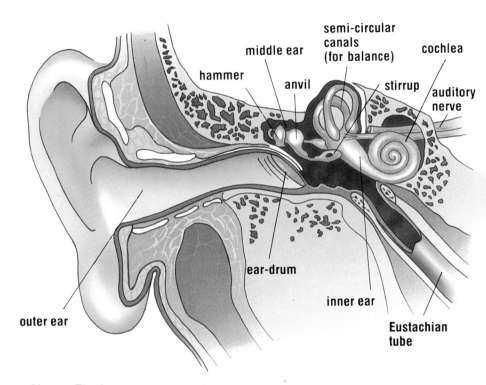

semi-circular canals (for balance)

middle ear

hammer

anvil

cochlea

stirrup

auditory nerve

ear-drum

inner ear

outer ear

Eustachian tube

Above: The human ear. Sound waves in the air make the eardrum vibrate, and the tiny bones called the hammer, anvil, and stirrup pass on the vibrations to the inner ear. There, inside the cochlea, the vibrations are converted into electrical impulses that travel to the brain.

Right: In general, animals have better hearing than humans. With its huge ears, the desert fox has exceptional hearing.

After sight, hearing is the most important of our senses. Our ears pick up sounds traveling through the air. But what exactly is sound? Stretch a rubber band over the mouth of a glass, pluck it, and it will twang. It is this vibration that gives rise to the twanging sound.

Sound is made up of vibrations. The vibrating rubber band, for example, vibrates the molecules of air in contact with it. These molecules in turn cause the molecules in contact with them to vibrate. These vibrations are passed on molecule by molecule, until eventually they reach our ears.

In the ear, the vibrating air molecules hit the eardrum and make it vibrate. Tiny bones pass on the vibrations to the inner ear, which sends "hearing" messages to the brain.

Sound moves away from a vibrating object in a kind of wave motion, just like waves move across the water when you throw a stone into a pond. Just like all waves, sound waves can vary in amplitude (height of wave) and frequency (rate of vibration).

Sound waves with a small amplitude are soft, while those with

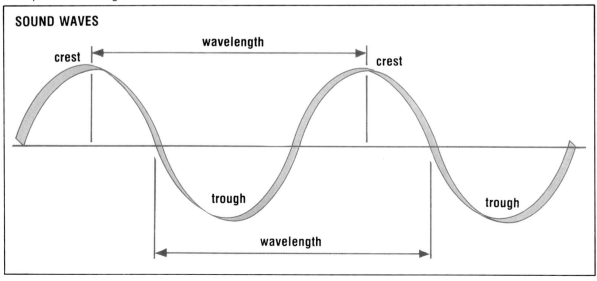

SOUND WAVES

wavelength

crest

crest

trough

trough

wavelength

Left: Sound travels in the form of a wave. It is a pressure wave. The crests are regions of high pressure; the troughs are low-pressure regions. The distance between two crests, or two troughs, is the wavelength. The frequency is the number of complete waves that pass a particular point in a second.

a large amplitude are loud. Those with a low frequency are low-pitched, like the voice of a bass singer; while those with a high frequency are high-pitched, like the voice of a soprano.

Some of the sounds that reach our ears are rhythmic and pleasant, like most musical sounds. The various types of musical instruments employ different means of setting up sound vibrations. Guitars and violins use vibrating strings and are called string instruments. Trumpets and flutes use vibrating air columns and are called wind instruments. Drums and cymbals vibrate when they are struck and are called percussion instruments.

Other sounds that reach our ears are irregular vibrations and are unpleasant. We call them noise. A jackhammer is one of the worst offenders. Its noise is so loud that people who operate it wear special ear protectors to protect their ears, which could be damaged by the noise level. People who work outside at airports also have to protect their ears. The roar which a jet makes is one of the noisiest things imaginable.

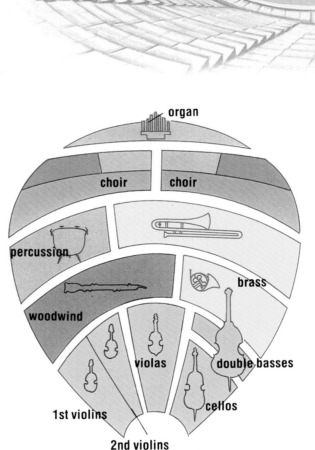

Above: The Greeks and Romans built tiered ampitheaters in which to perform plays. These structures had remarkable sound qualities or acoustics.

Left: The various sections of a symphony orchestra are often laid out in this way. The strings are in front, then come the woodwind, brass, and percussion. Choir and organ, if any, are located at the rear.

MAKE A GUITAR

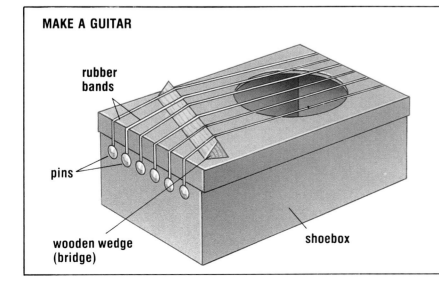

You can make a simple, guitar-like instrument out of rubber bands and a shoebox. Cut a hole in the top of the box as shown. Stretch rubber bands over the lid and pin them in place. Make some quite loose and others quite taut. Push a wedge of wood under them so that they are clear of the lid. Play your guitar by twanging the bands. Adjust their tension as necessary to "tune" them into a musical scale.

SOUND RECORDING

Thanks to modern methods of sound reproduction we can listen to the songs and music of our favorite singers and composers in our own homes, even in the street, with the same clarity we would hear in the concert hall.

The original means of recording sound was devised by the American inventor, Thomas A. Edison, in 1877. He used a vibrating needle to record sound in, and play it back from, grooves in a wax cylinder. From his original "phonograph" evolved the present-day record player, which preserves sounds in the grooves of plastic disks.

As in any sound reproduction system, the recording process begins with a microphone. It converts sound waves into varying electrical signals. In disk recording, these signals make a needle, or stylus, cut wavy grooves in a rotating lacquered disk. From this, commercial disks, or records, can be produced.

On a record player, a disk is rotated on a turntable, and a stylus picks up vibrations from the grooves. Crystals, or other devices, convert the vibrations into electrical signals, which are fed to a loudspeaker. And this reproduces the sound.

The latest disk players, however, use lasers to play back recorded sound, from small compact disks. Sound is recorded on them in the form of microscopic pits. The sequence of pits and flats represents the waves of sound very precisely in a digital code of 0s and 1s. The laser reads the code very precisely, so that the sound is reproduced very accurately.

Today, the most popular means of sound reproduction, however, is magnetic tape. The availability of small portable cassette players, like the Sony Walkman, means that people can play their favorite music anytime, anywhere.

In tape recording, the varying electrical signals from a microphone are sent to an electromagnet, where they set up varying magnetic signals. These signals create a distinctive pattern in the magnetic particles on the tape. On playback, the reverse happens: the tape runs past another electromagnet, this senses the magnetic pattern and converts it back to electrical signals, which are then fed into a loudspeaker.

Above: A group "cutting an album" in a recording studio. The microphones feed signals into a complicated electronic machine called a mixer, whose operator adjusts them to produce a balanced sound which is recorded on magnetic tape. This may then be used to cut a master disk, from which commercial records are produced.

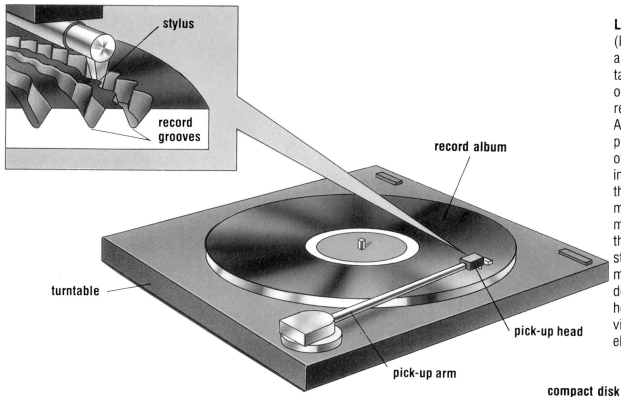

stylus

record grooves

record album

turntable

pick-up head

pick-up arm

Left: Playing an LP (long-playing) record on a turntable. The turntable rotates at a speed of exactly 33.3 revolutions per minute. A stylus (needle) in the pick-up head at the end of the pick-up arm rests in the spiral groove of the record. The movement of the record makes the wavy sides of the groove vibrate the stylus. An electromagnetic or crystal device in the pick-up head converts these vibrations into variable electrical signals.

Right: The latest disks, called compact disks, are much smaller than albums. They measure less than 5 inches across. On these disks, sounds are recorded in a pattern of microscopic pits etched in the surface. A laser beam is used to "read" the pit pattern and generate the signals that are fed to a loudspeaker.

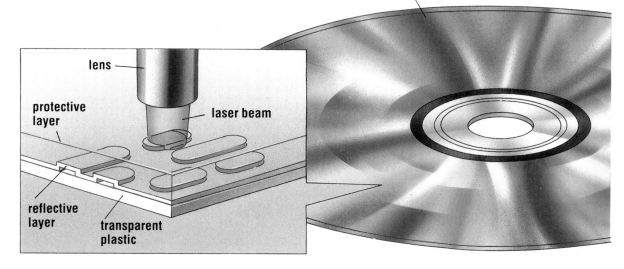

compact disk

lens

laser beam

protective layer

reflective layer

transparent plastic

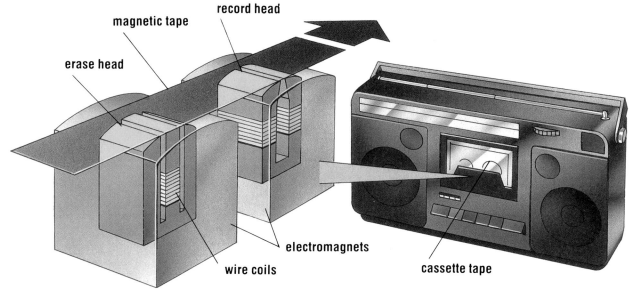

record head

magnetic tape

erase head

electromagnets

wire coils

cassette tape

Left: Cassette tape recorders and players use magnetic tape wound on reels in a self-contained box, or cassette. When recording, the tape passes over two "heads," which are small electromagnets. Amplified microphone signals are fed into the coils of the record head. There they set up equivalent magnetic signals, which imprint a magnetic pattern on the tape as it moves past.

SILENT SOUND

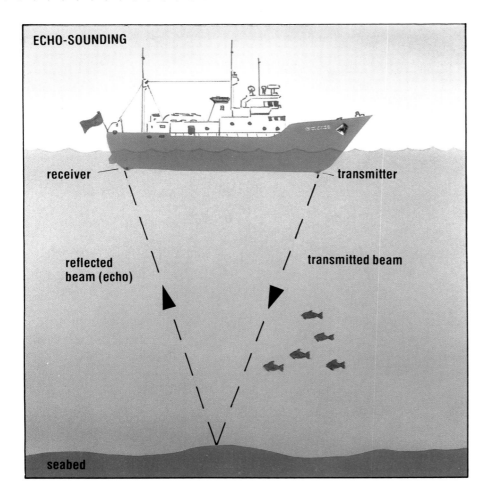

ECHO-SOUNDING

receiver

transmitter

reflected beam (echo)

transmitted beam

seabed

Above: Ships and boats use sound waves to sound, or find the depth of, the water they are sailing in. They are equipped with echo-sounders, which bounce beams of ultrasonic waves off the seabed and measure the time they take to travel between the transmitter and receiver. Sound waves travel faster through water than they do through air. In air at sea level, the speed of sound is about 750 mph. In the sea, it is nearly 3,400 mph.

Small bats, like the horseshoe bat, have poor eyesight. Yet, when they go out hunting at dusk, they can home in on moths and other insects with uncanny accuracy. This is because they locate their prey using sound waves. They emit sound waves and listen with their sensitive ears for any echoes, which may tell them that a meal is nearby. In the sea, dolphins use a similar echolocation method to help them find fish.

We cannot hear bats or dolphins hunting, however, because the sound waves they emit have a higher frequency than our ears can detect. We call such sounds ultrasonic. Human ears can detect sound waves with a frequency, or rate of vibration, up to about 20,000 Hz/sec (cycles/sec). Some dolphins, however, emit sound waves with a frequency of up to 200,000 Hz/sec.

Scientists, engineers, ships navigators, fishermen, and doctors have found ultrasonic waves very

Above: A bat homes in on its insect prey, emitting high-pitched squeaks and listening for their echoes. For this kind of echo-location, bats use sound with a frequency of up to 120,000 hertz (cycles/second). This is way beyond the range of human hearing.

useful. For example, fishermen emulate the dolphin and use echolocation to find shoals of fish. Their boats are equipped with echosounders, which send out ultrasonic waves and listen for echoes. This is a kind of underwater radar, which called sonar. It is also used more generally by boats for finding water depth. The time between the transmission of waves, and the receipt of the echoes from the seabed, is a measure of the distance they have traveled, and of the depth.

Doctors use a similar method to look inside the human body. Their sonar scanners send out ultrasonic waves, which are reflected by body tissues and organs. From the pattern of reflections, a computer can build up an image of the internal structure. Scanning is used most widely to picture an unborn baby in its mother's womb. Sonar scanning is much safer than taking X-rays because sound waves are harmless, whereas X-rays may be harmful in large doses.

Below: A pregnant woman undergoes a sonar scan in hospital to examine the baby developing in her womb. A probe sends out pulses of ultrasonic waves and picks up their echoes.

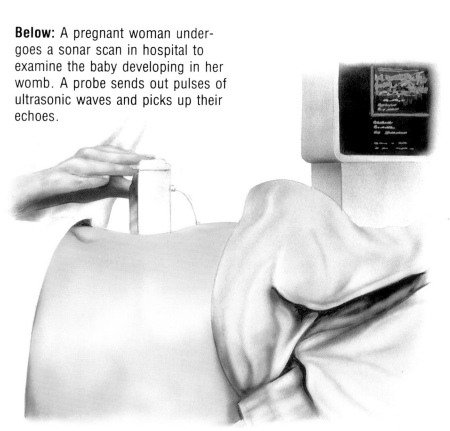

Below right: Signals picked up during sonar scanning of a pregnant woman are computer-processed and displayed on a video screen. This sonar scan, in false color, shows the head and shoulders of a seven-month-old fetus.

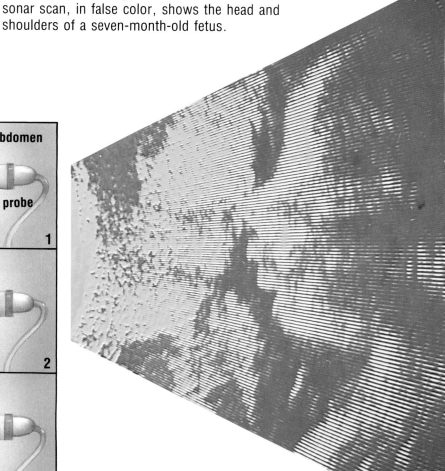

Right: This series of pictures shows how the sonar scanning of a pregnant woman works. (**1**) The ultrasonic probe is passed over the surface of the abdomen, and a pulse of ultrasonic waves enters the body. (**2**) When the pulse hits the wall of the womb, part of it is reflected, producing an "echo", which the probe picks up. (**3**) The pulse penetrates further until it reaches the baby. Part is again reflected, producing another echo.

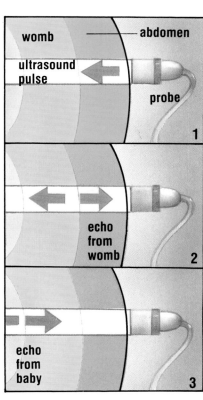

womb — abdomen
ultrasound pulse
probe
1

echo from womb
2

echo from baby
3

INDEX

Figures in **bold** refer to captions.

A

acid rain **16**, 17
acids 17
actinides **18**
adhesives 30
air
    in atmosphere **15**
    cold 43
    expansion of 43
    force of moving **32**
    heated **42**, 43
    as matter 9
    as mixture of gases 14
air currents **42**, 43
aircraft 51, **51**
alloys 24, **25**
aluminum 20, 23
amalgams **25**
ammonia 23
amphitheaters **71**
amplitude 70, 71
amplitude modulation (AM) **58**
Andromeda galaxy **36**
animals, and hearing **70**
antibiotics **30**, 31, **31**
antifreezes 30
argon 14
aspirin 30
Atlantis space shuttle orbiter **37**
atmosphere 14, **15**, **39**
atomic number **18**, 20
atoms 20, **20**
    and crystal shapes 11
    and elements **18**
    and lasers **68**, 69
    and lighting **52**
    and nuclear power 47, **47**
Aurora Borealis **39**

B

Baekeland, Leo 28
Bakelite 28
balloons, hot-air 14, **14**
bats 74, **74**
batteries 54, **54**
Bell, Alexander Graham 56
binary code 61, **61**
biotechnology 31
biotite **11**
blast furnaces 24, **24**, **25**
boiling point 9, 26, 27
bronze 24, **25**
burning 17

C

calcite **10**, 11
calculators **60**
cameras 65
    TV 59, **59**
capillarity 13
carbon **10**, **16**, 17, 19, 24, 26, 28,
    45, 54, **54**
carbon dioxide 14, 17
carbon monoxide 24
CERN (European Center for
    Nuclear Research) 21

chlorine 14, 30
circuit boards **60**
coal, 17, **23**, **44**, 45
cobalt 38
coinage 24
coke 24, **24**
colors 66, **66**, **67**, **68**
compact disks 68, 72, **73**
compass 38
compounds 19, 28
computers 61, **61**
conduction 43
convection 43
copper 19
CPU (central processing unit) **61**
cracking 27, 28
crop-spraying 30
crude oil 23, 26, **26**, **27**
crystals 10-11, **10**, **11**

D

detergents 27, 30
diamonds **10**, 11, **11**
distillation 26, 27, **27**
drugs 27, 30, 30-31, **31**
dynamo effect 55

E

ears 70, **70**, 71, 74
Earth
    and geothermal energy 49
    and gravity 35, **36**
    and magnetism 38, **39**
    and matter 9
echolocation 74, **74**, 75
echo-sounders 74, 75
Edison, Thomas A. 72
electric charges 20, 21, 53, 57
electric current 38, 39, 54-7, **54**
electricity
    current 54-5, **54**
    and magnetism 38
    mains 54-5, **55**
    and nuclear power plants 47
    static 52, 53, **53**
    and water power 48, **48**
    and wind power 49, **49**
electromagnets **39**, 54, 55, 56, 57,
    72, **73**
Electromagnetic spectrum 66, 67
electromagnetism 38, 55
electrons 20, **20**, 54, 59
elements **18**, 19, 20
energy
    alternative 48
    and burning 17
    chemical 43
    electrical 43
    geothermal 49
    heat 43
    kinetic 43
    light **68**, 69
    nuclear 47, **47**
    and particles of matter 9
    solar **48**, 49
engines
    diesel 50, 51
    jet 51, **51**

petrol 50-51, **50**
erosion 16, 17
ethene/ethylene 28
exosphere **15**
eyes **64**, 65

F

fax (facsimile machine) 57, **57**
feldspar 10, **11**
fireworks **17**
First Law of Motion 32
fluids 12
fossil fuels 44-5, 48
fractionation 26, 27, **27**
Freedom space station **37**
frequency 70, 71
frequency modulation (FM) **58**
friction **32**, 33, 53
fungicides 30

G

galaxies, and gravity 35, **36**
galena **10**, 11
galvanizing **11**
gamma rays 66, **67**
gas 44, 45, **45**, 47
gases
    density 14
    invisibility of 14
    particles turn into 9
gems 11
generators 55, **55**
genetic engineering 31
geodes 11
geostationary orbit 37
glass 23
glowworms **16**
gluons 21
gold 19, **25**
granite 10, **11**
graphite **10**
gravitons 21
gravity 33, **33**, **34**, 35, 36, **36**, **37**,
    38, 43
"Green Revolution" 30
"greenhouse" gas 14
guitar, home-made **71**

H

heat **42**, 43
heating 9
helium 14
herbicides 30
Hiroshima 47
holography 68-9, **69**
hovercraft 51
hydraulic pressure **13**
hydrocarbons 26, 27, 30
hydroelectric schemes 48, **48**
hydrogen 14, 19, 26
hydrogen sulfide 14

I

ice
    in polar regions **16**
    water turns to 9, 16
icebergs **16**
igneous rock, **8**, **11**

inertia 32
infrared rays 66
insecticides **27**, 30
insulators 53
Intelsats 37
ionosphere **15**
iron 23, 24, 38
iron ore 24, **24**
iron oxide 17

K

Kennedy Space Center **37**
kinetic theory of matter 9

L

lanthanides **18**
lasers
    light 68-9, **68**
    and telecommunication **56**
LCD (liquid crystal display) **60**
lead sulfide **10**
lenses 63, **63**, **64**, 65
levers 40, **40**, **41**
light 62-3, **62**, **64**, 65, **65**, 66, **66**,
    67
    laser 68-9, **68**, **69**
lightning **52**, 53
lignite **23**
limestone 24, **24**
liquids
    and melting point 9
    properties of 12-13
lodestone 38
LPs (long-playing records) **73**

M

machines 40-41, **40**
magnetism 38, **38**, **39**
Marconi, Guglielmo 58
mass **34**, 35
matter 9, 35, 38
melting point 9
Mendeleyev, Dmitri **18**
Meteosat 37
mica 10
microphones 57, 58, **58**, 72, **72**, **73**
microscopes **64**, 65
microwaves 58, **58**, 66
minerals
    and carbon dioxide 17
    crystals 10, **10**, 11, **11**
    and mining 23
mining **22**, 23, **23**, **44**
mirrors 62, **62**, **63**, 65, **68**, 69
molecules 20, 28, 43, 70
Moon
    and gravity 35
    and laser beams **68**
Morse, Samuel 56
Morse Code 56
musical instruments 71, **71**

N

neutrinos 21
neutrons 20, **20**, **47**
Newton, Isaac 32, 33, 34
Niagara Falls **11**
nickel 38

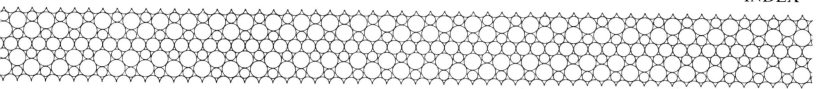

nitrogen 14, 23
Northern Lights **39**
nuclear fission 47, **47**
nuclear fusion **47**
nuclear power 47, **47**
nucleus **18**, 20, **20**
nylon 28, 30

O
oil 44, 45, **45**, 47
oil refineries 26, **26**, 27, **27**, 28
ore **22**, 23, 24, **24**
organic chemistry 26
oxidation 17
oxygen **10**
   and atmosphere **15**
   as basic building block 19
   and coke 24
   and fossil fuels 45
   and rusting 16, 17

P
parachutists, free-fall **35**
particle accelerators 20, **21**
penicillin 31
Periodic Table **18**, 20
periscopes **62**
pesticides 30, 45
petrochemicals 27
petroleum 23, 26, 27, 30
pharmaceuticals **27**, 30
phonographs 72
photons **21**
pig iron 24, **25**
pistons 50, **50**, 51
pivots 40, **40**, 41
plastics 23, **27**, 28, **28**, 45
pneumonia 31
pollution 17
polyethylene 28
polymerization 27, 28
polymers 28
power stations
   coal-fired **55**
   nuclear 47
precious stones 11
prisms 66, **66**
protons **18**, 20, **20**
pulleys 41
PVC (polyvinyl chloride) 28

Q
quarks **21**
quartz 10-11, **10**, **11**

R
radiation 43, 47
radio 58
radio waves **15**, 58, **58**, 59, 66
rainbows 66, **66**
RAM (random-access memory)
   61
record players 72
records 72, **72**, **73**
reflection 62, **62**, **67**, **68**, 75
refraction 63, **63**, 66
rock
   and carbon dioxide 17

changing state of **8**, **9**, **11**
   and crystals 10
   and fossil fuels 44, **45**
   igneous **8**, **11**
ROM (read-only memory) **61**
ruby 11
rusting 11, 17, **16**

S
salt 23
sapphire 11
satellites 36-7, **36**, **56**, 58, **58**
screws 41
sea 23
ships, and ultrasonic waves 74-5,
   74
silica **10**
silicon chips **60**, 61
silver 19
skaters **12**, 13
slag 24, **24**
smelting 24
sodium chloride 23
solar energy **48**, 49
solids
   and particles of matter 9
   properties of 12
sonar scanning 75, **75**
sound
   recording 72, **72**
   silent 74-5, **74**, **75**
   varieties of 70-71, **70**, **71**
sound waves 70-71, **70**, 72, 74,
   74, 75
space shuttles **37**
space stations **37**
spacecraft
   and exosphere **15**
   and gravity 36, **36**
spectrum 66, **66**
Sputnik 1 36
stars
   and energy **47**
   and gravity 35, **36**
   and light 65
   and matter 9
steel **11**, 24
   stainless **25**
steelmaking 24, **25**
stratosphere **15**
styrofoam 28
subatomic particle 20, **21**
sugar 10
sulfur **10**, 23
sulfur dioxide 14, 17
sulfuric acid 17, 20
Sun
   "greenhouse" effect 14
   heat of **42**, 43, **43**, **48**, 49
   and light 62
   and matter 9
surface tension 13
surgery, and lasers 68, **69**
synthetic fibers **27**, **29**, 45

T
tape recording 72, **72**, **73**
telegraph 56, **56**

telephone 56-7, **56**, **57**, 58, 68
telescopes 65, **65**
television 59, **59**
telex 56
temperature 9, 12, **13**, 43, **43**
terminal velocity **35**
tetanus **31**
thermals **42**
thermometers 12, **13**
thermoplastics 28, **29**
thermosets 28
Third Law of Motion 33
transformers **55**
troposphere **15**
turbines 49, **49**, 50, 51, **51**, 55, **55**
turbofans 51
turbojets **51**
turboprops 51

U
ultrasonic waves 74-5, **74**, **75**
ultraviolet rays **43**, 66
universe, and gravity 35
uranium 47, **47**

V
Van Allen radiation belts **39**
VDUs (visual display units) 61,
   61
vibrations 57, 66, 68, 70, **70**, 71,

   72, **73**, 74
vidicons **59**
volcanic eruptions **8**, 9
voltage 55, **55**

W
water
   changing state of 9
   as commonest liquid **11**
   as a simple compound 19
   "skin" on **12**, 13
water cycle 16
water power 48
water vapor 9, 14, 16
waterwheels 48, **48**
wavelengths 62, 66, **66**, **67**, 68,
   68, **70**
weight **34**, 35
wheel and axle **40**, 41
wind 32, 33, 43
wind power 49
wind turbines 49, **49**
windmills 49, **49**
word processors 61

X
X-rays 66, 75

Z
zinc **11**, 54, **54**

ACKNOWLEDGMENTS

ILLUSTRATIONS
Garden Studios Graham Austin 70 (centre), 74 (bottom)
Roger Courthold 20-21, 23, 27, 28-29, 38-39, 44-45, 46-47, 50-51, 62-63
Art Beat Richard Dunn 8-9, 12, 16-17, 30-31, 32-33, 74-75
Maltings Partnership 10-11, 13, 14-15, 18-19, 24-25, 40-41, 54-55, 56-57, 58-59, 60-61, 64-65, 70-71, 73
Garden Studios Darren Pattenden 34-35, 36, 42-43, 52-53
Art Beat Ian Thompson 48-49, 66-67, 68-69

Canon (UK) Ltd. 57, for permission to use reference material

PHOTOGRAPHS
AEA Technology 46 (top and bottom)
J. Allan Cash Ltd. 13, 14, 15, 16 (top), 17 (top), 26, 32, 33, 45, 51, 57, 68
The Bridgeman Art Library/National Trust 34
British Coal Corporation 44
Bruce Coleman Ltd. Keith Gunnar 22; P. A. Hinchliffe 17 (bottom); Norbert Rosing 39; Kim Taylor 12
Geoscience Features Picture Library Dr. B. Booth 11 (top and bottom)
Robin Kerrod 36, 37 (top and bottom), 65
Panos Pictures Alain le Garsmeur 16 (bottom)
Redferns 72
Science Photo Library CERN 21 (top), CNRI 30, 75, Mike Devlin 25 (bottom right), Simon Fraser/Northumbria Circuits 60, Simon Fraser/Pharmacy Dept., Royal Victoria Infirmary, Newcastle 31, Patrice Loiez 21 (bottom), Omikron 64, Philippe Plailly 69 (top)
John Watney 25 (top and bottom left)
Zefa Picture Library (UK) Ltd. T. Ives 52, Tom Tracy 69 (bottom), Simon Warner 66